INFERTILITY

Philosophy and Medicine

VOLUME 53

Editors

H. T. Engelhardt, Jr., *Baylor College of Medicine, Houston, Texas*

S. F. Spicker, *Baylor College of Medicine, Houston, Texas*

CATHOLIC STUDIES IN BIOETHICS 3

Founding Editors

Francesc Abel, S.J., *Institut Borja de Bioethica, Barcelona, Spain*

John C. Harvey, *Center for Clinical Bioethics, Georgetown University, Washington, D.C.*

Series Editor

Kevin Wm. Wildes, S.J. *Department of Philosophy, and the Kennedy Institute of Ethics, Georgetown University*

Advisory Board

Paolo Cattorini, *Istituto Scientifico, Milan, Italy*

Bernard Hoose, *Heythrop College, University of London, London, UK*

Johannes Huber, *University of Vienna, Vienna, Austria*

James F. Keenan, S.J., *Weston School of Theology, Cambridge, Massachusetts*

James J. McCartney, O.S.A., *Villanova University, Villanova, Pennsylvania*

Jean Porter, *University of Notre Dame, Notre Dame, Indiana*

Paul J.M. van Tongeren, *Catholic University, Nijmegen, The Netherlands*

The titles published in this series are listed at the end of this volume.

INFERTILITY

A Crossroad of Faith, Medicine, and Technology

Edited by

KEVIN Wm. WILDES, S.J.

Assistant Professor of Philosophy, Senior Research Scholar of the Kennedy Institute of Ethics, and the Center for Clinical Bioethics, Georgetown University, Washington D.C., U.S.A.

KLUWER ACADEMIC PUBLISHERS

DORDRECHT / BOSTON / LONDON

Library of Congress Cataloging-in-Publication Data

```
Infertility : a crossroad of faith, medicine, and technology / edited
 by Kevin Wm. Wildes.
     p.   cm. -- (Philosophy and medicine ; v. 53)
   Includes bibliographical references and index.
   ISBN 0-7923-4061-2 (alk. paper)
   1. Infertility--Religious aspects--Catholic Church.  2. Human
 reproductive technology--Religious aspects--Catholic Church.
 3. Catholic Church--Doctrines.  I. Wildes, Kevin Wm. (Kevin
 William), 1954-     . II. Series.
 RC889.I557  1996
 176--dc20                                              96-16497
```

ISBN 0-7923-4061-2

Published by Kluwer Academic Publishers,
P.O. Box 17, 3300 AA Dordrecht, The Netherlands

Kluwer Academic Publishers incorporates the publishing
programmes of D. Reidel, Martinus Nijhoff,
Dr W. Junk and MTP Press.

Sold and distributed in the U.S.A. and Canada
by Kluwer Academic Publishers,
101 Philip Drive, Norwell, MA 02061, U.S.A.

In all other countries, sold and distributed
by Kluwer Academic Publishers Group,
P.O. Box 322, 3300 AH Dordrecht, The Netherlands

Printed on acid-free paper

All rights reserved

© 1997 Kluwer Academic Publishers
No part of the material protected by this copyright notice may be reproduced or
utilized in any form or by any means, electronic or mechanical,
including photocopying, recording or by any information storage and retrieval system,
without written permission from the copyright owner.

Printed in the Netherlands

TABLE OF CONTENTS

PREFACE   vii

KEVIN WM. WILDES, S.J. / Technology, Reproduction, and Faith   1

SECTION ONE: SCIENTIFIC AND PSYCHOLOGICAL QUESTIONS

JAMES A. SIMON / Advances in Assisted Reproductive Technologies   9

YULA PONTICAS & PETER J. FAGAN / Issues in the Psychological Evaluation and Care of *In Vitro* Fertilization Couples   27

MIRIAM B. ROSENTHAL / Therapy of Working With the Childless Woman: The Pathos of Unrealized Dreams, The Psychology of Female Infertility   39

SECTION TWO: THEOLOGICAL REFLECTIONS

JAMES F. KEENAN, S.J. / Moral Horizons in Health Care: Reproductive Technologies and Catholic Identity   53

WILLIAM E. MAY / *Donum Vitae*: Catholic Teaching Concerning Homologous *In Vitro* Fertilization   73

JEAN PORTER / Human Need and Natural Law   93

JOHN W. CARLSON / Interventions Upon Gametes In Assisting the Conjugal Act Toward Fertilization   107

CAROL A. TAUER / *Donum Vitae*: Dissenting Opinions on the "Simple Case" of *In Vitro* Fertilization   125

SECTION THREE: MORAL AND SOCIAL REFLECTIONS

DEBORAH D. BLAKE / Infertile Couples: Psychological
    Needs, Social Responsibilities   149
PATRICIA BEATTIE JUNG / What Price Fertility?   167
KEVIN WM. WILDES, S.J. / *In Vitro* Fertilization: Secular
    Moral Authority, Biomedicine, and the Role of the State   181

SECTION FOUR: CONTINUING CONVERSATION

RON HAMEL / Epilogue   197

APPENDIX: SACRED CONGREGATION FOR THE DOCTRINE
    OF THE FAITH / Instruction on Respect for Human Life
    in Its Origin and on the Dignity of Procreation: Replies
    to Certain Questions of the Day   209

NOTES ON CONTRIBUTORS   239

INDEX   241

PREFACE

As with other books in the *Catholic Studies* sub-series of the *Philosophy of Medicine* series this volume is part of a conversation among Catholic theologians, philosophers, and moralists. It is a conversation informed by the scientific work of men and women in infertility medicine and research as well as by counselors and psychiatrists who care for infertility patients. The discussions could not have been possible without the support of The Steering Committee on Bioethics of the International Federation of Catholic Universities and generous foundational support. John Collins Harvey and Francesc Abel, S.J. were crucial to this conversation in their leadership of the initial conference with which this volume began. Francesc Abel, S.J. has served as an extraordinary leader of the Steering Committee in his ability to identify and address issues in bioethics that are of great importance to the Church, society, and medicine. Our debt to him is enormous.

The most recent form of this conversation, this volume, could not have taken place without the work and patience of the authors in this volume. A special note of thanks is due to Angela Jacobson and Linda Wiederhold for their help in preparing this manuscript. For everyone's patience and cooperation, especially the authors, I am most grateful.

Georgetown University Kevin Wm. Wildes, S.J.
Washington, D.C.

KEVIN WM. WILDES, S.J.

TECHNOLOGY, REPRODUCTION, AND FAITH

Technology can change everything. Technology opens horizons and makes possible what was once, at best, only imaginable. In so doing technologies can change human practices and, in altering human practices, technologies have the potential to influence and change moral values. While neither good or bad in itself technology changes the way men and women see the world and act in it. In so doing technology can influence what is identified as appropriate and inappropriate moral acts and values.

In the modern era technology often has been aimed more and more at production, development, and manipulation. It has made old tasks easier and more efficient while also opening new possibilities. Medicine has been affected more than most areas of human life by technological development. Technological applications of medical knowledge have given medicine the ability to actually do things to change the course of disease or illness. For centuries there was very little that medicine actually could do to heal for patients. However, with the development of the research science in the nineteenth and twentieth centuries medicine increased its ability both to diagnose and treat patients. Since medical technologies have been disseminated widely the effects on medical practices and values have also been widespread.

Because there is a profoundly moral quality to medicine the use of medical technologies has opened new moral discussions and questions. Medicine touches the lives of men and women at crucial moments in a person's moral narrative. The expanded horizon of medical choices brought about by medical technology represents a challenge to moral values and assumptions. Biological life can be extend and death can be postponed, birth can be controlled, and some of the most basic human characteristics can be altered. The reproductive technologies of the last two decades offer wonderful opportunities to childless couples. At the same time these technologies raise profound moral challenges to couples and traditional Christian morality. These technologies challenge traditional assumptions about parenting, family, the roles of men and women.

In the last decade there has been vigorous discussion within the Roman Catholic Church about the use of these new reproductive technologies. In 1987 the Congregation for the Doctrine of the Faith issued *Donum Vitae* (Instruction on Respect For Human Life In Its Origin And On The Dignity Of Procreation, see Appendix). The document spoke to a variety of moral issues in reproductive technology such as the use of donor sperm and egg or the practice of surrogate mothering. *Donum Vitae* also addressed the use of reproductive technologies in "the simple case" between a husband and a wife with no outside parties involved as donors. DV argued that, in most cases, the use of reproductive technologies was wrong. In many cases (e.g. surrogate mothering and the use of donor sperm and egg) DV argues that the technologies undercut the very nature of sexuality, family, and marriage. Rather than arguing that these technologies posed grave moral dangers in the simple case DV maintained that the practice was wrong insofar as it created a split between the two ends of marriage: unity of the couple and procreation. The substance of this teaching relied on lines of thought developed earlier by Pope Pius XII (see Pius) has been reaffirmed in subsequent statements by John Paul II (see John Paul). The teachings have been the subject of widespread discussion in the life of the Church (see, Wallace and Hilgers; McCormick; Shannon and Cahill; Verspieren).

The essays in this volume offer continued reflection on the simple case, DV, and some of the related issues in reproductive medicine. However, as these essays make clear, the simple case is far from simple in that it raises significant questions for the secular culture as well as for the Roman Catholic community. Even the simple case raises important psychological, social, and moral questons. The simple case is embedded in assumptions about family, sexuality, economics, and society

Reproduction is not merely a biological event. It is a cultural, moral, social, and psychological event. In the opening essay of this book J. Simon reviews the important biological background of infertility as well as the history of *in vitro fertilization* (IVF) and the different methods involved. Two accompanying essays in the first section, one by Yula Ponticas and Peter Fagan and the other by Miriam Rosenthal, examine some of the psychological issues associated with infertility and the uses of IVF. Children play a significant role in the understanding men and women have of their own lives (Ponticas and Fagan). The importance of children leads to pressures to have children which in turn lead to the great costs of addressing infertility. The costs come in terms of money, psychological stress, and careers. Ponticas and Fagan raise important questions about how a couple might respond to the stress of failure of IVF. Rosenthal argues that the desire for a child is psychological, biological, as well as social. She, like Ponticas and Fagan, points out that much of human identity is shaped around assumptions and experience of family and there can be profound grief reactions for an infertile couple. These reactions can be accompanied by anger, guilt, or blame. The inability to conceive or carry to term may be interpreted by couples as personal failure or deficiency. Rosenthal's essay also compliments Simon's essay insofar as she investigates the psychological causes of infertility.

But the psychological and biological issues are not the only questions addressed by the teachings of the Church. Sexuality and procreation are viewed as activities with moral significance. That is, they are concerned with who we are, what we are, and what we ought to be as human beings. The essays in the second section of the book turn to examine the theological teachings of DV. The essays by James Keenan, William May, Jean Porter, and Carol Tauer examine the theological arguments and methodology set out in DV. John Carlson examines a number of unresolved issues in the aftermath of DV.

Keenan's essay focuses on methodological considerations in evaluating DV method of theological argument. In Keenan's analysis DV's moral methodology of sexuality and parenting is too narrowly focused on the physical act. He argues that the focus of the moral evaluation on the *physical act* is not traditional in Roman Catholic thought. Rather, the moral tradition has focused not on the physical act but on the role of the intention of the agent. May offers a sympathetic reading of DV and its central thesis about the inseparability of the unitive and procreative ends of marriage. He argues that there are three elements – union, procreation, and marriage – which ought to form an unbreakable union. He argues that theology of the body, underlying DV, supports procreation not reproduction. Jean Porter's essay challenges the inseparability thesis that is so central to the arguments of DV. Porter argues that DV's interpretation of inseparability is too physical and fails to take the existential, relational understanding of inseparability into account. John Carlson's essay reviews questions surrounding certain reproductive technologies (e.g., GIFT and TOTS) in light of DV. Carlson raises important questions about why certain technologies and procedures are seen as acceptable while others are not.

Carol Tauer's essay raises both methodological and contentful issues in response to DV and the development of reproductive technologies. Tauer is critical of DV insofar as it frequently makes assertions following principles but fails to make argument for either. The document, she argues, does not address the problem of interpretation and application in that it is not always clear as to how principles are to be applied. Tauer picks up the questions of the social imperatives on biological parenting that are often imposed by society. Social expectations and demands often fail to raise issues about the exploitation of women or the uses of health care resources for these technologies.

The broader social themes are further examined in the third section of the volume in essays by Deborah Blake and Patricia Jung. Blake argues that the Roman Catholic tradition has viewed always within the social context. The use of resources for health care is an issue, in part, of the common good. She points out that reproductive technology deploys a great deal of resources. Jung takes up the question of justice and the use of resources. She points out that in the allocation of resources a society must ask whether or not infertility is a disease. The final essay in the section, by Kevin Wildes, turns toward the section of DV which calls for legislative action. While the document has many important concerns that a society may wish to legislate DV does not take into account the difficulties for such moral legislation in a secular, morally diverse society.

In a final essay in the volume Ron Hammel reviews the essays as a whole and raises important questions for the position of the Congregation as well as for the authors in this volume. His questions point towards the future direction of this discussion. The essays in this volume are an example of a tradition in development. The technological developments in reproduction that have raised real moral challenges to understandings of family and sexuality to which the Church has tried to respond. In DV the hierarchical Church responds to these challenges by articulating through how technologies ought to be understood and used. The essays in this volume are a response to DV and a further step in the ongoing conversation of these issues. The Church's response to such complex moral issues is itself complex and comes from different sources. One central source is the hierarchical church while another source has been the discussion among the community of theologians and moralists ... It is in the respectful dialogue that the tradition will develop and continue to respond to God's Word and human experience ...

BIBLIOGRAPHY

John Paul II: 1994, *Evangelium Vitae*, found in *Orgins* 24, April 6, 1994, United States Catholic Conference, Washington, D.C.
McCormick, R.A.: 1989, *The Critical Calling: Reflections on Moral Dilemmas Since Vatican II*, Georgetown University Press, Washington, D.C.
Pius XII: 1956, 'Address to the second world congress on fertility and sterility', *Acta Apostolica Sedes* **48**, 467–474.
Shannon, T.A., and Cahill, L.S.: 1988, *Religion and Artificial Reproduction: An Inquiry into the Vatican 'Instruction on Respect for Human Life'*, Crossroad, New York.
Wallace, M. and Hilgers, T.W.: 1990, *The Gift of Life: The Proceedings of a National Conference on the Vatican Instruction on Reproductive Ethics and Technology*, Pope Paul VI Institute Press, Omaha
Verspieren, P.: 1987, 'Les fecondations artificielles. A propos de l'Instruction romaine sur le don de la vie', *Estudes* **336**, 615–632.

SECTION ONE

SCIENTIFIC AND PSYCHOLOGICAL QUESTIONS

JAMES A. SIMON

ADVANCES IN ASSISTED REPRODUCTIVE TECHNOLOGIES

I. INTRODUCTION

Louise Brown, the world's first "test-tube" baby, burst into the public consciousness in 1978. As the description suggests, she was conceived with the assistance of reproductive technology. Since 1978, such technology – the means by which fertility problems in both men and women are often solved – has advanced dramatically, permitting more infertile couples than ever the ability to choose to have their own child.

Today, physicians have at their fingertips procedures which range from the commonplace to the extraordinary. Because advances in reproductive technology are so rapid, procedures considered experimental only a few years ago are now performed routinely. Others, though perhaps thought experimental today, are likely to become "standard of care" in a relatively short time. Regardless of the level of acceptance, all reproductive technology shares the distinction of having the same goal – the birth of a baby.

This article explains and examines various methods of assisting the reproductive process utilizing recent developments in technology, and offers insights into cutting-edge technologies that can help infertile couples, both now and in the future. Each of the methods discussed expand or enhance *in vitro* fertilization – the technology by which Louise Brown was conceived. These methods include the use of hormones to facilitate the growth of eggs in women with ovulation problems, transcervical intrafallopian transfer, cryopreservation, and the use of donor oocytes and embryos.

Additionally, investigators continue work in the areas of micromanipulation of an egg to allow substandard sperm to enter and fertilize it, and the ability to diagnose embryonic defects before the embryo is implanted. Although the ethical and legal issues surrounding this technology are serious and difficult, as we will discuss, they raise no question sufficient to outweigh the substantial benefit of providing potential parents the assistance they need to have a biological child – if that is their choice.

II. A BRIEF REVIEW OF REPRODUCTIVE PHYSIOLOGY

In order to gain insight into the technology available to assist infertile couples, one must also have a basic grasp of the essentials of reproductive physiology. An understanding of the intricate steps that must occur to allow the sperm and egg to successfully develop into a viable embryo, is crucial to the ability to visualize the truly remarkable nature of these innovations.

In the first step in the reproductive cycle, ovulation, one oocyte or ovum in an ovary matures and is released with the rupture of an ovarian follicle. Fertilization then occurs when a sperm and egg meet, usually in the oviduct (fallopian tube). In fertilization, the sperm penetrates the surrounding shell – or zona pellucida – of the egg, and the genetic material (DNA) from the sperm and egg unite. The new cell, called a zygote, immediately begins dividing, first into two cells, then four, then eight, and so on, as this early embryo passes down the oviduct and into the uterus. The embryo is now at the "blastocyst" stage.

During the next step, the embryo hatches out of the zona pellucida, attaches to the lining of the uterus, and begins to implant. In order for this early pregnancy to continue, it must be maintained by increased production of ovarian and placental hormones. Increasing amounts of progesterone come from the corpus luteum, which develops in the ovary at the site of the ruptured follicle. Human chorionic gonadotropin (hCG), secreted by the implanting embryo, increases, further stimulating the corpus luteum. Ultimately, the placenta itself begins to produce its own hormonal support, including progesterone.

In order to simulate, replace or assist the processes described above in the case of a couple whose reproductive system can not adequately perform these functions, scientists in the early seventies perfected the process though which Louise Brown and thousands of other children have been conceived, known as *in vitro* fertilization and embryo transfer, ("IVF-ET").

III. IN VITRO FERTILIZATION

A. The History of IVF

The birth of Louise Brown in 1978 marked the beginning of a new era in the treatment of infertility. Brown was the first baby conceived outside of the human body using the then radical and highly controversial technique.

The development of IVF-ET, now shortened to IVF, was the result of many different medical advances made in the 1960s and 1970s. In the 1960s, the advent of the radioimmunoassay (RIA) enabled researchers to detect and measure minute amounts of reproductive hormones accurately and rapidly. This advance permitted characterization of reproductive health and disease. Contraceptive and fertility researchers made use of this new technology throughout the 1960s and 1970s in the identification and purification of numerous natural and synthetic reproductive hormones. These developments gave rise to the clinical treatment of infertility. Building on these advances, infertility researchers used microsurgery, fiber optics, and ultrasound – which for the first time allowed visualization and retrieval of eggs – to develop IVF and other novel reproductive technologies.

The use of the highly sophisticated IVF treatment requires obtaining mature eggs (oocytes) from a woman's ovaries and transferring them to a culture dish. Next, the physician adds sperm from the woman's partner. Once fertilization occurs and the fertilized eggs have begun to divide, two or more of the resulting embryos are transferred to the woman's uterus.

In conventional IVF, which is virtually unchanged since the conception of Louise Brown, oocytes are obtained through a surgical procedure such as laparoscopy. (Although the use of surgical procedures is still common, such use is rapidly giving way to newer, non-surgical – and thus safer – techniques for collecting eggs.) The mature oocytes obtained can be the result of natural ovulatory cycles. Generally, however, ovulation induction with fertility drugs, such as clomiphene citrate, human menopausal gonadotropins (hMG), gonadotropin releasing hormone (GnRH) analogues or agonists, alone or in combination, is preferred because these drugs produce many more mature oocytes than the one or two commonly produced during a natural ovulatory cycle. Development of oocyte-containing follicles in the ovaries is monitored by ultrasound measurements and blood estradiol levels. When the follicles are judged mature, the eggs are retrieved by needle aspiration of the fluid inside the follicle with the enclosed ovum during the laparoscopy. In conventional IVF, transfer of the embryo into the woman's uterus is accomplished transcervically in the physician's office. After transfer, progesterone or human chorionic gonadotropin (hCG) may be given to supplement the luteal phase hormonal environment in an attempt to increase the likelihood that the embryo will implant.

B. Tubal Damage and IVF

For some infertile women, a major impediment in the normal progression of the reproductive process develops when there is damage to the fallopian tubes, and possibly to the ovaries – often caused by pelvic inflammatory disease. Pelvic inflammatory disease is usually the result of the sexually transmitted diseases chlamydia and gonorrhea. Surgery can often correct tubal blockage or other damage to the fallopian tubes (i.e., endometriosis), and the rate of success for such surgery has been greatly improved by the implementation of improved microsurgery techniques. However, this surgery is effective only for about half the women who have it performed. Approximately 50% remain infertile even after surgery. For women who cannot be helped surgically, and for those with other causes of infertility, hope of becoming pregnant remains in large part because of the development of assisted reproductive technologies such as IVF and its successor innovations.

IV. AN OVERVIEW OF THE NEWER REPRODUCTIVE TECHNOLOGIES

Ongoing research in the field of infertility treatment continues to spawn new technologies to enhance IVF and other methods of assisted reproduction. These technologies may provide hope for those men and women who were not helped by conventional treatments. For those couples with severe reproductive problems, these breakthrough methods may be their only chance for the conception and birth of a baby. While many new treatments are being researched to assist couples who are trying to overcome reproductive problems, several approaches under investigation appear to offer the most hope – new fertility drugs, egg retrieval/transfer methods, freezing techniques, donation regimens, and micromanipulation, including assisted hatching and preimplantation genetic diagnosis. A brief description of each of these procedures and processes follows.

GnRH Agonists for Controlled Ovarian Hyperstimulation. Protocols for IVF programs usually include the use of fertility drugs to achieve ovarian hyperstimulation and ovulation induction. These drugs – clomiphene citrate, human menopausal gonadotropins, and gonadotropin releasing hormone agonists – produce many more mature oocytes than the one or two commonly produced in a natural ovulatory cycle.

In the last four to five years, the use of gonadotropin releasing hormone (GnRH) agonists have become widely used as adjuncts to gonadotropins for ovarian hyperstimulation (Kubik et al., 1990, pp. 836–841; Benadiva et al., 1990, pp. 479–485; Lindner et al., 1990, pp. 140–144). GnRH agonists are compounds structurally related to endogenous GnRH, but chemically modified to assist proteolytic metabolism. Because these compounds are often thousands of times as potent as the native GnRH compound, their injection results in markedly increased hormone production. The agonist first stimulates the pituitary cells to secrete large amounts of the gonadotropins FSH (follicle stimulation hormone) and LH (luteinizing hormone) leading to multiple follicle development. However, there is a delayed secondary action – suppression of the release of further FSH and LH. This secondary action allows the physician to better control the maturity and number of viable oocytes ready for aspiration or ovulation and overriding the body's natural tendency to ovulate spontaneously only one or at most two oocytes at a time.

These agonists were initially administered for approximately three weeks ("long protocol") to completely suppress production of the woman's own ovulatory hormones, allowing more precise control over hormone levels and development of more mature follicles. More recently, IVF ovarian hyperstimulation regimens have called for shorter administrations of GnRH agonists ("short protocols") for *stimulation* rather than suppression of the woman's own natural FSH and LH secretion. In these "short" protocols the GnRH agonist is administered either near the end of the ovulatory cycle preceding the cycle from which the eggs will be harvested (luteal start) or coincident with gonadotropin stimulation at the start of the egg retrieval cycle (follicular start). These regimes not only recruit more follicles, but also improve ovarian hyperstimulation in some poor responders. Additionally, in women who ovulate spontaneously or prematurely when treated with gonadotropins, it is possible to achieve better cycle control.

Transvaginal Oocyte Recovery. As noted earlier, another recent, but widely adopted, development in IVF is the nonsurgical retrieval of mature eggs from the ovaries using ultrasound guided transvaginal ovarian puncture (Dellenbach et al., 1988, pp. 111–124; Wikland et al., 1988, pp. 103–110; Barak et al., 1988, pp. 585–588). With this technique, a needle is used to aspirate eggs from mature ovarian follicles by guiding it through the posterior wall of the vagina using transvaginal ultrasound guidance. Transvaginal oocyte recovery is rapidly replacing the more invasive and costly laparoscopic aspiration of ovarian follicles, not only because it is generally better tolerated by patients, but also because it avoids the need for general anesthesia and hospitalization.

Natural Cycle IVF. A third variation on traditional IVF techniques is to retrieve eggs during a woman's natural cycle rather than during a cycle in which the ovaries have been hyperstimulated with hormone (fertility drug) injections. Although, natural cycle IVF produces fewer embryos for implantation into the uterus, this may be offset by an improved implantation rate by avoiding the possible negative effects of hyperstimulation on the uterine lining and embryo receptivity. By avoiding ovarian hyperstimulation, natural cycle IVF is less costly and does not carry the risk of medical complications that in rare cases can result from ovarian hyperstimulation. In addition, natural cycle IVF avoids many of the ethical concerns related to cryopreservation and/or donation of "excess" oocytes or embryos. In addition to these positive factors, pregnancy rates achieved with this technique have been only slightly lower than for stimulated cycle IVF (Paulson et al., 1992, pp. 290–293).

Transcervical GIFT/ZIFT. For infertile women whose fallopian tubes are normal, gamete intrafallopian transfer (GIFT) and zygote intrafallopian transfer (ZIFT) are attractive techniques for achieving pregnancy when less invasive methods have failed. In GIFT, sperm and eggs are placed directly into the fallopian tube. In ZIFT, the egg and sperm are combined *in vitro* and the fertilized ovum, known as a zygote, is placed into the fallopian tube. The use of ZIFT or GIFT appears to yield higher pregnancy rates than IVF, although this finding remains controversial. Research indicates that the difference may occur because the fallopian tube offers a better environment for the development of the preimplantation embryo during its first few cell divisions than either the laboratory culture dish or the recipient uterus.

Previously, the placement of the gametes or zygotes into the fallopian tube could only be accomplished by laparoscopy under general anaesthesia. Now, however, these techniques can be performed nonsurgically by passing a fine catheter through the vagina, cervix, and uterus, using ultrasound to guide it into the fallopian tubes (Jansen et al., 1988, pp. 288–291) for placement of either sperm and eggs (GIFT) or embryos (ZIFT). This technique is similar to transcervical embryo transfer commonly used to deliver embryos into the uterus for IVF except that ultrasound is needed to negotiate the uterotubal junction. Use of both these nonsurgical methods (oocyte retrieval and transfer) in transcervical GIFT or ZIFT allows these treatment processes to be done entirely in the physician's office. these "outpatient" services result in significant financial savings.

Cryopreservation. Scientists have studied the effects of very low temperatures for over two hundred years, beginning with early experiments by Spallanzani in 1776. Fertility specialists have used freezing techniques to preserve sperm for artificial insemination since the early 1970s. In the 1970s the practice of cryobanking of human semen grew in the United States, spurred in part by the desires of some men to obtain "fertility insurance" before undergoing vasectomy. Significant numbers of live births produced using frozen sperm were reported by the mid-1960s. Today, freezing techniques are being used to preserve both eggs and embryos, as well as sperm, thus providing both men and women the same "fertility insurance". Cryopreservation allows either partner to provide viable gametes to the assisted reproductive process at some future time. For example, the man whose testicles must be removed because of cancer and the woman who is at high risk of familial ovarian cancer could equally insure against losing the ability to participate in conception by storing frozen gametes or embryos.

In recent years recognition of the risk of transmitting diseases, particularly the virus that causes AIDS, has greatly increased the use of frozen rather than fresh semen for artificial insemination. The quarantining of semen for at least six months is recommended by the American Society For Reproductive Medicine (ASRM). In its 1987 survey the Office of Technology Assessment (OTA) of the U.S. Congress found that frozen semen was used by 78 percent of U.S. physicians who did artificial inseminations. Although studies have shown higher pregnancy rates with the use of fresh semen, use of frozen sperm offers the maximum guarantee of avoiding the transmission of sexually transmitted diseases. It also allows for genetic screening of donor semen, although the OTA survey found that less than one-half of U.S. physicians did such screening.

Today, assisted reproduction programs are routinely using cryopreservation of embryos because this allows multiple cycles of IVF or ZIFT to be performed following a single ovarian hyperstimulation and egg retrieval procedure. Freezing unfertilized eggs also offers advantages for assisted reproductive technologies such as IVF, GIFT, or ZIFT. Freezing eggs retrieved during an IVF or GIFT cycle could allow multiple cycles without requiring additional drug stimulation and egg retrieval. In addition, oocyte freezing avoids many of the ethical issues of embryo freezing since the frozen oocyte is a single cell and can not develop into a human being without fertilization by sperm. However, to date, cryopreservation of oocytes (mature eggs), while attempted many times, has met with little success. Experimental data on oocyte freezing is relatively scarce and only three pregnancies have been recorded using frozen eggs (Pensis et al., 1989, pp. 787–794; Trounson, 1990, pp. 695–708; Levran et al., 1990, pp. 1153–1156).

Rather than using frozen oocytes, and in spite of the ethical implications, IVF programs around the world are increasingly using cryopreservation of embryos. The most recent data available from the United States IVF Registry shows that in 1990 there were 3,290 frozen embryo transfer cycles performed in 129 clinics. These transfers resulted in 382 pregnancies and 291 births for a success rate of 9 percent. Two clinics accounted for 20 percent of the pregnancies and 19 percent of the births. There were 23,865 embryos frozen through IVF cycles.

Much more widely researched than egg cryopreservation, freezing of embryos offers the same advantage of allowing multiple cycles of IVF or GIFT following a single ovarian hyperstimulation and egg retrieval cycle. Cryopreservation also offers the ability to reduce the risk of multiple pregnancies by transferring only one or two embryos at a time, though this approach also reduces the overall chances of any pregnancy in that cycle. Though widely used, this technique raises many more ethical issues than freezing of gametes, particularly regarding what should be done with embryos remaining after a couple's desired pregnancies have been achieved, after the couple has died, or if a marriage has ended in separation or divorce.

Co-Culture Techniques. It is widely recognized that the poor viability of *in vitro* fertilized and cultured embryos are largely responsible for the low rate of implantation and subsequent successful full-term births using IVF-ET (10–15%). Research suggests that one of the major causes of reduced embryo viability is the actual process of *in vitro* fertilization and subsequent culture of the embryo in the synthetic tissue culture medium. *In vitro* fertilization and culture yield embryos which are retarded in *rate* of growth relative to *in vivo* fertilized and cultured embryos. One attractive approach to improving the culture system used for fertilization and embryo growth is to co-culture embryos with other cell types, such as cells of the reproductive tract, which may provide a stimulus for embryonic development and health (Bongso et al., 1990, pp. 893–900) thereby simulating the nurturing environment of the fallopian tube (*in vivo*).

Research on defining the critical nutrients or growth factors that may be provided to the embryo by such co-cultured cells has identified embryo-derived platelet activating factor (PAF) as one possibility (O'Neill et al., 1989, pp. 769–772). Recent work has shown that supplementation of the culture media for human embryos with PAF results in a dramatic increase in the cultured embryos' developmental and pregnancy potential (Bongso et al., 1990, pp. 893–900). Other factors and secretory products of a variety of cell types are under intensive investigation.

Epididymal Sperm Aspiration. Male infertility can be caused by a lack of production of viable sperm or by blockage of the male genital tract caused by congenital abnormalities, inflammation, or surgery (vasectomy). For men infertile due to blockage, epididymal sperm aspiration, a new surgical technique for removing sperm from the obstructed reproductive tract. This approach can be an effective treatment (Asch and Silber, 1991, pp. 101–110). Researchers have demonstrated that viable sperm can be retrieved using this technique and successful pregnancies can then be achieved using this sperm for IVF. The potential candidates for this new treatment are many. Although estimates vary, a significant proportion of the 2 to 4 million infertile couples in the United States are unable to conceive because of male infertility. Some researchers have estimated that in nearly one-quarter of infertile married couples the husband has an abnormally low sperm count in the ejaculate. It has been estimated that sperm is completely absent from the ejaculate (azoospermia) of some 345,000 married men of whom some 40,000 suffer from a congenital absence of the vas deferens, the condition most amenable to the epididymal sperm aspiration process. In addition to these 40,000 men, candidates for sperm aspiration would include the tens of thousands and perhaps hundreds of thousands of men who have had an unsuccessful attempt at vasectomy reversal.

Donor Oocytes and Embryos. The use of donor oocytes and embryos is an obvious outgrowth of IVF (Navot et al., 1986, pp. 513–524; Sauer et al., 1990, pp. 1157–60). Egg or embryo donation involves the transfer of an oocyte or embryo from a fertile donor to an infertile recipient. Donor eggs can be transferred using the GIFT procedure (fertilized *in vivo* or they can be fertilized *in vitro*, as in IVF. *In vitro*, they can be fertilized using the recipient partner sperm or by donor sperm. All these variations are well known today and thousands of healthy babies have already been born using these techniques. The use of donated eggs or embryos to alleviate infertility in women is relatively new, but a modification of the technology has been used in animals for more than a century (Heape, 1890).

The procedure usually begins with gonadotropin hormone stimulation of the ovaries of the woman donating the egg or embryo. This is done to increase the number of mature eggs produced in one cycle. In egg donation, mature eggs are recovered from the donor and fertilized *in vitro* using the sperm of the infertile patients' partner. The resulting embryo is then transferred into the recipient. Following the century-old paradigm established by Heape, embryo donation, more recently called ovum transfer, uses a slightly different approach. In embryo donation the donor is inseminated to achieve *in vivo* fertilization, usually during her natural cycle. After the embryo reaches the donor's uterus (typically on the seventh day after ovulation/insemination, the embryo is recovered from the donor by non-surgical uterine lavage. The recovered embryo is then transferred to the recipient transcervically. All of these procedures are performed in an office setting without anesthesia (Bustillo et al., 1984a, pp. 889; Bustillo et al., 1984b, pp. 1117–1173; Buster et al., 1984, 53–60; Buster et al., 1985, 211–217). A key to successful egg or embryo donation is to ensure that the donor and the recipient are at the same stage (within two days) of the menstrual cycle. Dyssynchromy between the donor and recipient's menstrual stage, or between the developmental stage of the embryo and the endometrium will result in reduced or failed implantation. Alternatively, if both gametes are required, donor egg can be fertilized *in vitro* with donor sperm and the resulting embryo transferred to the recipient woman.

This technique makes it possible for women who lack ovaries, whose ovaries are not functional, or who have serious genetic abnormalities to become pregnant and have healthy children. Most women who request egg or embryo donation suffer from premature ovarian failure. This is a relatively common condition with menopause occurring before age 30. It affects about 4 percent of women.

The United States IVF Registry's most recent data show that in 1990, 67 clinics (37 percent) reported performing IVF with donated oocytes. There were 498 patients who received 547 donor transfers. The donor transfers produced 160 (29 percent) clinical pregnancies of which 122 (22 percent) resulted in live deliveries. Among these deliveries were 36 sets of twins, 3 sets of triplets, and 1 set of quadruplets.

Micromanipulation. Another promising new assisted reproduction technique is micromanipulation. Micromanipulation, akin to microscopic surgery, is currently used to enhance fertilization in the event of severe sperm dysfunction in the male partner, to assist "hatching" of the *in vitro* fertilized embryo in cases of recurrent failed implantation, and most recently as a method of preimplantation genetic diagnosis. is micromanipulation of either the embryo or the oocyte.

A. Types of Micromanipulation Techniques

To date, four microsurgical techniques have been used in fertilization efforts: zona drilling, partial zona dissection, subzonal insertion, and direct microinjection. In the first technique, zona drilling (ZD), the zona pellucida, the thick, transparent layer surrounding the oocyte, is not surgically pierced. Instead, small holes are "drilled" in the outside layer using an enzymatic medium that can digest this layer. Sperm can then pass through the holes to fertilize the egg. This method has been largely discontinued due to observations of abnormal oocyte and embryo structural changes (mostly polyspermies) along with inhibited growth, possibly due to the effect of the "drilling" enzymatic medium on the oocyte.

The most promising – and least invasive – method thus far has been partial zona dissection (PZD). In this technique, a microneedle pierces the zona pellucida to assist the passage of sperm into the egg. This approach has produced offspring in several species, including humans (Brinsden and Rainsbury, 1992, pp. 205–226). The major advantage of PZD is its minimal invasiveness relative to other micromanipulation methods. Disadvantages are that many sperm are required to accomplish fertilization, and that polyspermy (the penetration of the egg by several sperm) occurs with higher frequency than with some other methods. This technique may be indicated when a male has too few sperm, a reduction in sperm vitality, or immunologic or unexplained infertility.

With the third method, subzonal insertion (SZI), pronounced "SUZY", the sperm is injected just under the zona pellucida into the perivitelline space (PVS). The sperm then binds with the oocyte plasma membrane to begin fertilization. This technique can be applied in cases where men have extremely low sperm counts. This technique is more invasive than PZD but is significantly less invasive then microinjection of sperm into the egg, which pierces the egg's plasma membrane. Both animal and human pregnancies have been reported with SZI.

The fourth method, microinjection of the sperm directly into the ooplasm of the egg, has been used when the male of the couple has sperm with structural abnormalities that make them incapable of penetrating the egg or binding with the plasma membrane of the egg after penetration. This is the most invasive micromanipulation method, posing the greatest risk of structural damage to the oocyte. A rapidly increasing number of human and animal live births have resulted from this technique.

B. Uses of Micromanipulation

Micromanipulation of the oocyte can be used to treat otherwise intractable fertility disorders, particularly male-factor infertility. Severe sperm dysfunctions include a reduction in the vitality of the sperm, a very low sperm count, abnormal sperm morphology, or sperm that cannot penetrate the egg. Microsurgery is used to inject the sperm into the egg or to make an opening in the egg's thick, transparent outer covering, the zona pellucida, in order to permit the sperm to enter the egg. This procedure is typically performed during IVF.

Another use for micromanipulation in infertility treatment is in "assisted hatching." With this technique, a slit is placed in the outer layer of the embryo before placement in the uterus. Investigators believe this may help embryos implant in the uterus because many IVF embryos may be impaired in their ability to hatch through the zona pellucida or to overcome the frequent dyssynchrony between the IVF embryo's stage of development (typically 4–8 cell stage) and that of the endometrium. This situation commonly occurs because the endometrium is advanced in its maturation due to the exceptionally high estrogen levels ensuing from the development of more than one oocyte at a time, while the embryo(s) may be slightly retarded in development due to the imperfect *in vitro* culture conditions. Normally, only blastocysts consisting of 100 cells or more would be expected in the uterus at the time of implantation.

The most recent use of micromanipulation is for genetic testing for families at risk of inherited genetic diseases such as cystic fibrosis and sickle cell anemia. It is now possible to take a microbiopsy of cells from an early embryo allowing the diagnosis of genetic disease prior to implantation of the embryo during ovum transfer (Buster et al., 1992, pp. 17–26) or IVF (Handyside et al., 1992, pp. 905–909). In the process of "preimplantation genetic diagnosis," a single cell (blastomere) is taken from the preimplantation embryo fertilized *in vitro* usually when it reaches the eight-cell stage. The DNA of the cell is amplified using the polymerase chain reaction (PCR), which creates a sufficient quantity of the DNA to be analyzed. The embryos identified as not having the disease-carrying gene can then be selected for implantation into the mother's uterus. This technique can be used with genetic diseases, such as cystic fibrosis, where the genetic location of the abnormal/normal gene is known and the major mutation(s) have been identified. (Handyside et al., 1989, pp. 347–349)

The technique was used recently in Britain to achieve the birth of a healthy baby girl from parents who were both carriers for the cystic fibrosis gene (Handyside et al., 1992, pp. 905–909). Another use for micromanipulation is for X-linked, recessive genetic diseases. Typically, males are affected by these diseases, such as hemophilia, as the diseased genes are recessive and found on the X sex chromosome. For families with backgrounds of these "male" diseases, physicians can use micromanipulation to determine the sex of IVF embryos before implantation, and only implant female embryos.

Thus, while used only for diagnosis, the ultimate goal of micromanipulation work with genetic diseases is to treat and cure the gene defect. Therefore, the affected IVF embryos would not have to be selected out and discarded if they carry a diseased gene. Rather, the embryos would be treated and implanted in the uterus. In the future, abnormal genes will be exchanged for normal ones by "injecting" the normal genetic material into the affected embryo and "exchanging" it for the abnormal genetic material.

Micromanipulation is also being developed in a controversial area that could allow for the creation of genetically identical individuals through cloning. Preliminary research in animals shows that nuclei from multicellular embryos can be injected into unfertilized, enucleated eggs and progeny of the same genotype created. Research that could lead to the development of a cloning technique in the United States is not likely to be developed in the near future as ethical issues, such as the creation of genetically identical humans,(identical twins, triplets, etc). must first be addressed and resolved. However, this techniques offers extraordinary promise for creating large numbers of identical twin animals. These twins would allow definitive testing so as to investigate the nature/nurture dichotomy.

Oocyte Maturation In Vitro. Development of immature human eggs retrieved from unstimulated ovaries is a promising new technique that could prove to be of great benefit to donor oocyte programs. Initial studies indicate that unstimulated, immature eggs obtained by needle aspiration can be successfully matured in culture using mature follicular fluid (Cha et al., 1991, pp. 109–113). Preliminary studies have achieved successful pregnancies using this technique and researchers say this procedure combined with cryopreservation might make it possible to establish egg banks, similar to sperm banks, for oocyte donation. With further development of these new techniques it may be possible to use cadaver ovaries or parts of ovaries obtained during surgery for other donor oocytes.

V. THE ETHICAL IMPLICATIONS OF REPRODUCTIVE TECHNOLOGY

Today, the incredibly wide range of treatments now available to infertile couples permits many couples the irreplaceable option to choose to have biological child. However, the same treatment options that are acting to enhance fertility, also raises questions about parentage and other ethical implications of assisted conception.

Prior to the advent of high-tech fertility treatments, the answer to the question "who are a child's parents?" was straightforward and unambiguous. A child's parents were the man and women whose sexual union produced the child with the man contributing one-half the genetic material of the child through his sperm and the woman one-half through her egg.

Of course, there is much more to parentage than genes. For the mother, there is carrying the developing fetus for approximately ten months and then giving birth to the baby. For the father there is the role of supporting and caring for the mother and the baby during this time. For both parents, this is only the beginning.

Most people would agree that the most important aspect of parenting is actually raising the child. Certainly, loving, caring for, and supporting a child for perhaps 20 years requires much more of the parents than did the conception or even the gestation of the child.

It is evident that today's infertility treatment options mean that the conception of a child may necessarily include the participation of more than just one man and one woman. Previously, questions about who may claim to be a child's parents have only arisen in cases of adoption. For example, are the parents the adoptive couple who is lovingly raising the child – shaping his or her character, moral and religious beliefs, and intellectual development? Or are the parents his or her "biological" parents, the people from whom the child has derived his or her unique genetic makeup?

Much publicized in recent years have been custody cases involving surrogate mothers who were implanted with an infertile woman's fertilized embryos or inseminated with the sperm of the infertile woman's husband. In these cases, who is the true mother? Is the woman who carried and gave birth to the child its mother or is the mother the genetic source of the egg from which the child has developed? Use of donor eggs, donor sperm, and donor embryos to assist a couple in achieving pregnancy raises similar thought-provoking questions.

Unfortunately, perhaps, the complex and important ethical questions and issues result in legal entanglements and issues which must be addressed. Because ethical considerations such as these are highly political as well as legal, they will, de facto, be answered long after the medical technology has surpassed the lawyers and bioethicists. Therefore, it is vital that those physicians and scientists of us working in this area have thoroughly considered the legal and ethical problems contemporaneously with the development of the technology. It may be irresponsible of scientists not to address the ethical implications of their research, when those implications could, if not adequately resolved, withhold fully developed technology from the intended beneficiaries merely because the legal and "ethical" communities are often years behind the learning curve.

BIBLIOGRAPHY

Asch, R.H. and Silber, S.J.: 1991, 'Microsurgical epididymal sperm aspiration and assisted reproductive techniques', *Annals New York Academy of Sciences*, 101–110.
Barak, Y. et al.: 1988, 'The development of an efficient ambulatory *in vitro* fertiliztion (IVF) and embryo transfer (ET) program using ultrasonically guided oocyte retrieval', *Acta Obstetrica Gynecologica Scandinavia* **67**, 585–588.
Benadiva, C.A. et al.: 1990, 'Comparison of different regimens of a gonadotropin-releasing hormone analog during ovarian stimulation for *in vitro* fertilization', *Fertility and Sterility* **53**(3), 479–485.
Bongso, A. et al.: 1990, 'Co-cultures: Their relevance to assisted reproduction', *Human Reproduction* **5**, 893–900.
Brinsden, P.R. and Rainsbury, P.A. (eds.): 1992, *A Textbook of In Vitro Fertilization and Assisted Reproduction*, Parthenon Publishing, Park Ridge, New Jersey, Chapter 12, pp. 205–226.
Buster, J.E. et al.: 1992, 'Embryo micromanipulation in reproductive medicine', *Seminars in Reproductive Endocrinology* **10**, 17–26.
Buster, J.E. et al.: 1984, 'Biologic and morphologic development of donated human ova recovered by nonsurgical uterine lavage', *Transactions of the American Gynecological Obstet Soc* **3**, 53–60.
Buster, J.E. et al.: 1985, 'Biology and morphology of donated human ova recovered by non-surgical uterine lavage', *American Journal of Obstetrics and Gynecology* **153**, 211–217.
Bustillo, M. et al.: 1984a, 'Delivery of a healthy infant following nonsurgical ovum transfer', *Journal of the American Medical Association* **251**, 889.
Bustillo, M. et al.: 1984b, 'Non-surgical ovum transfer as a treatment in infertile women: Preliminary experience', *Journal of the American Medical Association* **251**, 1117–1173.
Cha, K.Y. et al.: 1991, 'Pregnancy after *in vitro* fertilization of human follicular oocytes collected from nonstimulated cycles, their culture *in vitro* and their transfer in a donor oocyte program', *Fertility and Sterility* **55**, 109–113.
Dellenbach, P. et al.: 1988, 'The transvaginal method for oocyte retrieval. An update on our experience (1984–1987)', *Annals New York Academy of Sciences* **541**, 111–124.
Handyside, A.H. et al.: 1992, 'Birth of a normal girl after *in vitro* fertilization and preimplantation diagnostic testing for cystic fibrosis', *The New England Journal of Medicine* **327**, 905–909.
Jansen, R.P.S. et al.: 1988, 'Nonoperative embryo transfer to the fallopian tube', *The New England Journal of Medicine* **319**, 288–291.

Kubik, C.J. et al.: 1990, 'Randomized, prospective trial of seuprolide acetate and conventional superovulation in first cycles of *in vitro* fertilization and gamete intrafallopian transfer', *Fertility and Sterility* **54**(5), 836–841,

Levran, D. et al.: 1990, 'Pregnancy potential of human oocytes – the effect of cryopreservation', *The New England Journal of Medicine* **323**, 1153–1156.

Lindner C. et al.: 1990, 'Ovarianresponse and pregnancy rates in *in vitro* fertilization, gamete intrafallopian transfer, and *in vivo* fertilization therapies after combined gonadotropin-releasing hormone agonist/human menopausal gonadotropin stimulation', *Gynecol Obstet Invest* **29**(2), 140–144.

Navot, D. et al.: 1986, 'Artificially induced endometrial cycles and establishment of pregnancies in the absence of ovaries', *The New England Journal of Medicine* **314**(13), 417–420.

O'Neill C. et al.: 1989, 'Supplementation of *in-vitro* fertilization culture medium with platelet activating factor', *Lancet* **2**, 769–772.

Paulson, R.J. et al.: 1992, '*In vitro* fertilization in unstimulated cycles: the university of southern California experience', *Fertility and Sterility* **57**, 290–293.

Pensis, M., Loumaye, E. and Psalti, I.: 1989, 'Screening of conditions for rapid freezing of human oocytes: Preliminary study toward their cryopreservation', *Fertility and Sterility* **52**, 787–794.

Sauer, M.V., Paulson, R.J. and Lobo, R.A.: 1990, 'A preliminary report on oocyte donation extending reproductive potential to women over 40', *The New England Journal of Medicine* **323**(17), 1157–1160.

Trounson, A.O.: 1990, 'Cryopreservation', *British Medical Bulletin* **46**, 695–708.

Wikland M. et al.: 1988, 'Oocyte retrieval under the guidance of a vaginal transducer', *Annals New York Academy of Sciences* **541**, 103–110.

YULA PONTICAS AND PETER J. FAGAN

ISSUES IN THE PSYCHOLOGICAL EVALUATION AND CARE OF *IN VITRO* FERTILIZATION COUPLES

I. INTRODUCTION

In vitro fertilization (IVF) was first employed in the late 1970's as an assisted reproductive technique available to infertile couples for whom conception has been difficult or impossible. The IVF technique is offered typically after the couple (most likely the woman) has undergone several years of diagnostic and failed treatment procedures. For most couples it is the last hope for biological parenthood. In addition to the stress of the chronicity of infertility and the procedures involved in the diagnosis and treatment, the couple as a unit and each individual may face a wide range of psychosocial and intrapsychic conflicts contributing to the burden of their infertility.

In the societal sense, the couple has to address emotionally laden issues raised by a culture in which parenthood is an integral and expected role of successful adulthood. Western culture is child oriented, the emphasis often being on "children are our future". This places a high value on having children as a way to contribute to society and to maximize personal fulfillment. Such pressure can filter down to a more personal level where infertile couples, particularly those who have chosen not to disclose their infertility, may be pressured by family members, co-workers or friends "to start a family".

Biological parenthood is a major, if not the major developmental task of adulthood. For infertile couples this task is obviously very difficult to achieve. Dealing with the often profound drive to procreate when biological limitations seriously thwart this drive is a significant challenge to each member of the couple and the couple as a unit. Often the wife will defer career goals because of the disruptive nature of the infertility problems and the often ill-based hope that she will become pregnant. Difficulty dealing with the very real possibility of biological childlessness can among other things produce a sense of inadequacy in gender and sex roles for one or both of the spouses (e.g., the threat of childlessness can highlight the discrepancy in desire for a biological child between the couple that otherwise may not have surfaced).

Several practical concerns can exist in a couple's quest for biological parenthood. The diagnostic and treatment procedures themselves may produce secondary sexual dysfunction in one or both of the partners (Fagan et al., 1986; Freeman et al., 1985; Morse et al., 1985). When the dysfunction occurs in the husband (e.g., erectile or ejaculatory problems) the existing pressure to perform sexually increases, anxiety heightens, and sexual functioning is further impaired, creating a vicious cycle. The couple may need to seek treatment for the sexual dysfunction, adding an additional sense of delay in the quest for conception and viable pregnancy as well as added expense.

In the similar vein, infertility treatments are costly and for many represent a serious financial drain. Many insurance companies do not reimburse for IVF procedures, or do so only partially. Particularly for the woman, career goals are often deferred due to the scheduling issues involved in infertility treatment. Procedures need to occur in often very narrow windows of time always subject to the menstrual cycle. Career commitments are hedged; income is decreased.

Infertile couples may need to deal with religious and ethical concerns before they make the decision to pursue IVF. For some couples (including Roman Catholics) the desire for a biological child supersedes any ecclesiastical prohibition while others actively struggle with their ethical concerns and seek the advice of clergy. Once the couple decides to stop medical intervention the religious and ethical beliefs may guide the couple in the decision to accept childlessness or to pursue adoption.

Infertile couples who are candidates for IVF typically are veterans of infertility treatments for whom the IVF procedure is the last hope and chance for biological parenthood. It is understandable that these couples present themselves in the best possible light for psychological evaluation prior to the initiation of the IVF procedure, particularly since the common perception is that the evaluation "qualifies" them for IVF. A competent clinician involved in a psychological evaluation with an IVF couple should attempt to achieve a balance between the social desirability factors that obfuscate problem areas and the knowledge of the psychological complexity of the infertility treatment process.

The John Hopkins Hospital (JHH) Division of Reproductive Endocrinology (Department of Obstetrics and Gynecology) has performed

IVF since 1984. In recent years approximately 60 couples per year elect this form of treatment with a "take home baby rate" of 18% as of 1990. Since 1989 psychological evaluation is a stated prerequisite of a couples' participation in the IVF process. The remainder of this paper will describe our method of evaluation with these couples with emphasis placed on the care of the couples as they go through the process. We will conclude with a few comments about the unitive and procreative meaning of the conjugal act as stipulated in *Instruction on Respect for Human Life in its Origin and on the Dignity of Procreation* (1987) in the context of our observations.

II. REFERRAL PROCESS

Couples who apply for consideration in the IVF program at the JHH receive an informational packet which explains the various steps of the IVF procedures in detail. It is in this packet that the patients learn that prior to their participation in the IVF program a psychological consultation is required. Although we presume that IVF patients are not psychiatric patients, the evaluation is presented to the couple as a forum for them to speak freely about their infertility experience, concerns they may have about the procedures, and about other stressors that may be present in their life. Psychological consultations are conducted by senior staff of our unit, the Sexual Behaviors Consultation Unit, which is an outpatient unit within the department of psychiatry of the JHH.

The reproductive endocrinologists requesting a psychological evaluation wish to know from us "how will the couple handle success or failure of IVF?". This is an especially relevant question for IVF candidates because, as mentioned previously, this is often a last stop or the last procedure that will give them hope of biological pregnancy. A couple is often coming face to face with their biological limitations and may in the course of IVF have to confront the possibility that they will not become biological parents. Because of these factors, normal reactions to stressful medical procedures such as IVF may be more pronounced and it may be helpful to the referring physician to have some information about how an individual may respond and if they are at risk for developing problems.

In attempting to answer the referral question as we proceed with an evaluation we employ four (4) criteria that serve as minimal exclusionary conditions. These are criteria we established early in our work with IVF couples (Fagan et al., 1986). These criteria do not select for "ideal" IVF candidates. We look for a minimal level of competence of each couple to integrate IVF into their lives in terms of the procedure itself as well as the results (continued infertility or pregnancy). The four criteria that we use are as follows (Fagan et al., 1986, p. 671):

1. *"Do the spouses give any evidence that their marriage is likely to dissolve in the near future?"* It is important to determine whether the hope of a child is being used to solidify a marriage that is in imminent danger of dissolution. Although this is a very unlikely scenario we want to avoid the stressor of simultaneous pregnancy and divorce as much as possible.

2. *"Is either spouse influenced primarily by the other or by a transient situational event to become a parent such that his/her authentic motivation can be doubted?"* The concept of "authentic motivation" is a difficult one to quantify. We like to insure that the desire for IVF is not a passing fancy or resulting pressure from a particular spouse. IVF, like gastric stapling, transsexual surgery, and penile implant surgery is a medical procedure that can radically alter an individual's way of life. We want to ensure that the individuals involved are very aware of all the ramifications and not responding to a superficial stimulus.

3. *"Is there a psychological condition present that would place either spouse at grave risk if attempts at IVF were unsuccessful?"* We look for particularly unrealistic expectations for the upcoming IVF that would result in a dramatic negative response should the procedure fail. In a related context if a woman has a history of miscarriage followed by severe depression we would want to be prepared for and examine very carefully with the patient how they would respond to a different sort of failure if IVF is not successful.

4. *Are there psychological conditions that would render the spouses incapable of assuming parental responsibilities?* Although it is not our responsibility to select who should or should not be parents, we do have the well being of the future child in mind as we evaluate prospective IVF candidates. We attempt to assess whether the couple has any condition that prevents them from providing adequate parenting. For example, if there is continuous substance abuse or an active major mental illness in either parent, this would point toward a poor prognosis for parenthood as long as the condition is present.

Again the above criteria serves as minimal conditions that must be excluded. There are no "ideal IVF candidates". We are looking to see how this couple will integrate this experience (often after years of having other reproductive failures) into their lives and how they will adjust to very likely continued infertility with the failure of IVF or with a pregnancy should the procedure succeed.

III. EVALUATION COMPONENTS

Our psychological evaluation for IVF candidates involves four components including couple interview, self-report psychometric tests, feedback to the couple, and letter of recommendation to the referring physician. Below is an elaboration upon each of these components:

1. *Interview.* The first and most primary component is the hour long interview with the couple. We seek to understand how the infertility experience has been for them as individuals and as a couple beginning with the initial diagnosis of infertility and how they each reacted to it. It is also helpful to assess each individual's understanding of the cause of the infertility. We go into some detail regarding the type of diagnostic and treatment procedures that the couple has undergone as well as their reactions, in particular we ask how the couple has handled infertility failures. For the woman, for example, having menstrual periods during months where she might have expected to become pregnant can be quite traumatic.

We spend time discussing the upcoming IVF procedure in terms of feelings about the procedure that each individuals may have as well as their expectations. IVF has a relatively low success rate ($<25\%$) and we want to make sure that the couple is not unrealistically optimistic to the point where their optimism exceeds the statistical probability and reality of success. Alternatively, we also try to get a sense of the extent of ambivalence about the upcoming IVF procedure and determine if one member of the couple wants the child more than the other. If there is no ambivalence, there may be possible denial of the negative components and the probable negative outcome of the procedure.

In discussing the IVF procedure we try to walk the couple through the two outcomes of the procedure. We ask about how they may handle a failed procedure and explain to them that there are many points throughout the procedure where a couple may drop out. We also talk about the possibility of success and of what a pregnancy and infant would mean to the couple. These questions provide us with a great deal of information and serve as an excellent way to engage the couple because they sense that we really want to hear from them about their experience.

The final purpose of the couples interview is to obtain abbreviated psychosocial histories from each individual. This would include educational, occupational and marital history as well as their family and personal psychiatric history, any history of substance abuse, and their impression of the quality of their marriage and concurrent stressors that may be present. We always ask about religion and how that has influenced their struggle with infertility. The final component of the psychosocial history includes a mental status examination for each individual. This is a cross sectional interview that seeks to identify any symptoms of psychiatric disorder especially affective illness, anxiety disorder or formal thought disorder that is ongoing in an individual.

2. *Self-Report Psychometric Data.* The psychometric instruments are mailed out to the couple several weeks prior to their scheduled appointment time and they are brought to the interview completed. Once scored, the data are incorporated into the evaluation report. We use the self-report psychometric data to corroborate, support and to enrich our interview findings. The following psychometric instruments are employed:

- *NEO Personality Inventory (NEO-PI):* This instrument assesses normal personality traits, and psychological well-being and coping styles in normal (i.e, non-psychiatric) populations (Costa and McCrea, 1985).
- *Brief Symptom Inventory (BSI):* This instrument measures current, acute psychological symptoms in the patient such as depression, anxiety, isolation (Derogatis and Melisaratos, 1983).
- *Michigan Alcohol Screening Test (MAST):* The MAST screens for past and present patterns of alcohol abuse (Selzer, 1971).
- *Dyadic Adjustment Scale (DAS):* The couples marital adjustment is measured in areas of marital satisfaction, consensus, cohesion, and affectional expression (Spanier, 1976).
- *Infertility Questionnaire:* This instrument looks for emotional impairment that can accompany infertility in areas of self esteem, blame, guilt, and sexuality (Bernstein, 1986).

3. *Feedback to Couple.* In the majority of the cases we are able to tell a couple at the end of the interview what our recommendation is as far as their continued participation in the *in vitro* procedure. At a later point the couple is welcomed to return for a session to review specific psychometric data. All couples are offerred copies of their evaluation report.

4. *Letter to Physician.* The letter to the referring physician accompanies a four to six page summary report of our evaluation. The three possible recommendations that we make to the referring physician are listed below:

1. To proceed with the IVF procedure with no psychiatric contraindications
2. To proceed with IVF while concurrently addressing a specific condition (e.g., anxiety disorder, mild depression, sexual dysfunction)
3. To defer IVF until a specific condition is treated (e.g., alcohol abuse or major depression). Typically this is handled by making a referral to a substance abuse or psychiatric professional.

Our experience with IVF couples has revealed individuals who have a history of a long, emotionally, draining course of infertility treatments. Perhaps as a testament to their tenacity and their ability to cope our patients have relatively normal personality profiles and in general have an absence of acute psychological symptomology such as anxiety or depression. There is little reported infertility specific impairment in terms of self-esteem, guilt, blame, or sexual functioning. The incidence of psychiatric disorder is comparable to the general population. There is relative absence of present alcohol abuse, and the marriages are generally quite stable with no more than minor conflict in areas such as money, in-laws, and time spent together. These findings support our initial assumption that IVF patients are non-psychiatric patients.

IV. CARE OF THE IVF CANDIDATES

Our involvement with IVF couples often does not stop after we have completed our evaluation. Although our evaluation is extensive and we consider it a therapeutic interaction in and of itself we are often involved with a given couple as they go through the IVF procedure if there is a need for consultation liaison work with the Division of Reproductive Endocrinology.

Although we believe in the therapeutic nature of the IVF evaluation we also believe it is helpful to prepare and support the couple as they go through the various stages of the IVF procedure. We typically advise couples as follows for each component of the IVF procedure:

1. *Induction.* Once a woman has made a decision to pursue *in vitro* fertilization, at the appropriate point of her menstrual cycle she is given hormones that will hyperstimulate her ovaries to produce the maximum amount of egg-containing follicles. It is important to inform the patients about how potentially disruptive this particular part of the procedure can be because for a two week period the woman needs to come into the hospital twice daily in the morning for blood drawing and sonograms and in the afternoon to receive an injection of the appropriate amount of hormones for that day. This can be very stressful when one has to juggle job and family responsibilities. Many women experience fatigue, headaches, and mood lability on the hormones. It is helpful to prepare both the woman and her husband for this sort of reaction.

2. *Retrieval.* Once the medical staff determine that ovulation is impending based on the woman's hormone levels and the size and number of her ovarian follicles, the eggs are retrieved via trans vaginal sonography and follicular aspiration. Women typically find themselves under a great deal of pressure to produce "good eggs" as the eggs are graded in quality. Eggs that do not meet the quality criteria are not fertilized. The male counterpart of the woman having to "produce eggs" is the actual fertilization which is completely dependent upon the man producing a semen sample through self-masturbation. The typical setting in which the man is asked to provide the sample is the bathroom. This very vital component of the IVF procedure places a great deal of pressure on the husband and can result in performance anxiety and sometimes the inability to ejaculate a semen sample. For both the husband and wife, going through the retrieval and fertilization components forces them to have their gametes examined microscopically, and have judgements made about them on a very basic, cellular level.

3. *Fertilization.* Once the eggs are fertilized and incubated for 48 hours, the embryos are placed in the woman's uterus. She then has to lie still

for four to six hours. Many women feel a great deal of responsibility during this part of the procedure. They are afraid to get up, to walk or go to the bathroom for fear they will somehow disrupt the implantation process. The women also receive large doses of hormones at this point to help facilitate the implantation. Sometimes, the side effects of the hormones mimic symptoms of early pregnancy. It is useful for the woman to be aware of this so as not to falsely raise expectations of pregnancy prematurely.

4. *Pregnancy Test.* Two weeks after the transfer the woman has two pregnancy tests several days apart from each other. For women who are already feeling pregnant, possibly because of the hormonal effects, a negative test is very emotionally distressing. Other women often will get their menstrual period with a similar reaction. There can be during this time an intense period of denial of the failure of the IVF procedure. On the other hand the sadness, anger and the emptiness of infertility is completely magnified when a negative pregnancy test is the result of the couples' last option for biological parenthood. There is often despair about what a couple should do next in terms of pursuing further IVF procedures, terminating any biological attempt or perhaps considering adoption.

After a failed IVF attempt, many choose to repeat the IVF procedure because the data suggest that the statistical probability of pregnancy increases with each time. Another reason couples tend to repeat is that even though the outcome may be a failure, on a given cycle much can be learned about a woman's reproductive physiology and her response to hormones and dosages needed as she goes through an IVF cycle. The procedure is "fine tuned" for the next time. Typically a couples' anxiety is somewhat diminished after they have gone through the procedure once because they are familiar with the routine. Their chances of success do increase statistically. They learn about the process of what they are going through and they achieve a greater sense of control over what is going on.

Some couples decide to terminate any attempts at having a biological child after an IVF attempt. This involves an intellectual as well as an emotional acknowledgement that biological pregnancy is not an option. This acknowledgement comes after a long period of waiting for a child they will never produce. Although the desire for a biological child may never disappear, we can help the couple place it in a perspective that allows them, especially the woman, to invest energy in other generative endeavors.

V. COMMENTARY ON UNITIVE VERSUS PROCREATIVE MEANING OF THE CONJUGAL ACT: PERSONAL REFLECTIONS

> A. It is in their bodies and through their bodies spouses consumate marriage and are able to become mother and father, the procreation of a person must be the fruit and the result of married love (*Instruction*, p. 27).

The IVF couple candidates that we see have a typically long history of repeated psychologically painful experiences in which their marriage is consummated but neither conception not parenthood is achieved. Bodies and souls are as involved in striving toward parenthood as couples who are blessed with children. Based on our data and observations, IVF couples have very stable loving marriages, relatively free of major conflicts. However, for reasons that are not always medically certain these individuals find it difficult, if not impossible, to conceive.

Becoming "mother and father" involves much more than the successful union of gametes in the woman's fallopian tubes. The IVF couples whom we have seen demonstrate great commitment to having a child together and have invested a great deal personally, emotionally, and financially to achieve what to the vast majority of couples is taken for granted. Their ability to have intercourse open to procreation remains intact but they are denied the desired outcome of a child. The women in particular subject themselves to numerous intrusive and often painful procedures only to endure with their husband the monthly grieving process when pregnancy is not achieved. Further, the couples often make sacrifices in other areas of their lives (e.g., financial, career) in order to pursue the possibility of biologic parenthood.

The process of infertility is often the greatest source of stress a marriage must endure. Yet the IVF couple candidates manage to continue responding out of their love for each other, viewing this challenge as yet another opportunity for their marital love and commitment to grow and their union to be strengthened. The vast majority of marriages blessed with children have not faced such a challenge, nor have felt the pathos, and emptiness of childlessness.

B. In homologous IVF/ET, therefore, even if it is considered in the context of "de facto" existing sexual relationships, the generation of the human person is objectively deprived of ots proper perfection: namely that of being the result and friut of a conjugal act in which the spouses can become "cooperators with God for giving life to a new person" (*Instruction*, p. 30).

We suggest that the "proper perfection" of the generation of the human person be inclusive of technical assistance. The maintenance of the human person's life and even his or her death will probably include technical biomedical assistance. Why not allow technical assistance a role at life's inception? The intrinsic factors in the proper perfection of human generation is sexual intercourse open to procreation and the presence of generative love in the parents. IVF couples' sexual expression is open to conception only with technical assistance. It is not the absence of technical assistance but its very presence which perfects the human generation. Unless one adheres to a hermeneutic of intercourse in which its physical components (penile ejaculation within the vagina) are more important than its spiritual components of generative and mutual love, one should be willing to "incorporate" technology into human generation.

This is what IVF couples do when they ambivalently turn to reproductive technology to assist them. They are grateful that the technology may make possible a conception and birth of a child. But they wish they did not have to use the technology. They employ the technology as they struggle to embrace each other in the difficult hours of the IVF process. The perfection of human generation occurs throughout the IVF in a moral unity which the couple struggles to maintain. As professionals involved in the care of IVF couples, we view our challenge as one of assisting them in maintaining that moral unity between their generative love and the difficult technological tasks they must complete to conceive.

BIBLIOGRAPHY

Bernstein, J: 1986: 'Infertility questionnaire', *Journal of Obstetrics, Gynecologic and Neonatal Nursing* **14**, 63–66.
Congregation for the Doctrine of the Faith: 1987, *Instruction on Respect for Human Life in Its Origin and in the Dignity of Procreation*, Vatican City, 1987.

Costa, P.T. and McCrae, R: 1985, *The NEO Personality Inventory Manual*, Psychological Resources, New York.

Derogatis, L.R. and Melisaratos, N: 1983, 'The brief symptom inventory', *Psychological Medicine* **13**, 595.

Fagan, P.J. et al.: 1986, 'Sexual functionings and psychosocial evaluation of *in vitro* fertilization couples', *Fertility and Sterility* **46**, 668–672.

Freeman, E.W. et al.: 1985, 'Psychological evaluation and support in a program of *in vitro* fertilization and embryo transfer', *Fertility and Sterility* **43**, 48.

Morse, C. and Dennerstein, L: 1985, 'Infertile couples entering an *in vitro* fertilization programme: A preliminary survey', *Journal of Psychosomatic Obstetrics and Gynaecology* **4**, 207.

Selzer, M.L.: 1971, 'The Michigan alcohol screening test: The quest for a new diagnostic instrument', *American Journal of Psychology* **127**, 1653.

Spanier, G.B.: 1976, 'Measuring dyadic adjustment: New scale for assessing the quality of marriage and similar dyads', *Journal of Marriage and the Family* **38**, 15–28.

MIRIAM B. ROSENTHAL

THERAPY OF WORKING WITH THE CHILDLESS WOMAN: THE PATHOS OF UNREALIZED DREAMS, THE PSYCHOLOGY OF FEMALE INFERTILITY

> And when Rachel saw that she bore Jacob no children, Rachel envied her sister; and she said unto Jacob: "Give me children or else I die".
>
> *Genesis 30:1*

The inability to bear children when one wishes is considered a developmental crisis in the lives of women and men. Throughout history, fertility has been highly valued and viewed as the link between generations. It is estimated that about ten to fifteen percent of couples of reproductive age in the United States are unable to conceive after a year of coitus without contraception. This represents about 3 million couples (Seibel, 1991) who are making over 2 million visits a year to the offices of physicians in their quest for a biologic child (Speroff, 1989). There does not seem to be an absolute increase in the occurrence of infertility (Center for Disease Control, 1985). Age adjusted rates remain about the same as they were in 1965, but many women are having their first children later in their thirties when fertility is declining. Peak fertility remains between the ages of 20 to 29 when many women today are building careers, and developing lasting relationships. Some other causes for infertility are the increase in sexually transmitted diseases and the possible presence of environmental toxins (Speroff, 1989). About 50% of couples' infertility problems are female, about 40% male, and the rest, problems in both partners.

The ability to reproduce has been seen by psychiatrists and psychologists as central to an individual's core gender identity, self-concept and body image, whether or not one wants children. Identifying with one's parents is part of adult development and part of that identification relates to the capacity to have children. Today's woman is presented with a wide variety of possible roles and choices, but she is still faced by a biological clock that limits her reproductive potential. This is not true for men who may remain able to reproduce into late adulthood. Sexual behavior is always influenced by the possibility of pregnancy. Most couples who are healthy, believe that when they want to become pregnant, they will be able to do so. Many have planned their lives carefully, have postponed marriage and childbearing while accomplishing other goals. There is considerable frustration, sadness and a feeling of being out of control when they learn that their plans are not being fulfilled.

Although the wish for pregnancy and a child are frequently part of the same process, they may not be. A woman may long for a pregnancy to be sure her body is working well, or if she has a chronic illness, that her reproductive organs are intact. The wish for a biologic child may be a response to parental, or societal pressures. It may reflect the concern for a relationship. It may occur in response to a loss or the death of a parent. It is frequently motivated by the longing to nurture a child, either to do a better job than ones parents or to do as well.

Burns (1993) has summarized the current psychological theories of infertility. While earlier writings focused on psychological causes for infertility especially that which was unexplained physically, more recent work has focused on the psychological distress caused by the inability to conceive and/or repeated pregnancy loss. Stanton and Dunkel-Schetter find that those aspects of infertility that couples find most stressful are unpredictability, negativity, uncontrollability and ambiguity. They use stress theory not only as a way of thinking about the psychological responses to infertility, but also as a way of trying to define what means of coping may be most helpful and successful to individuals and couples.

Burns herself has added to the stress theory the idea of boundary ambiguity, in which a family member may be psychologically present though physically absent. The wished for child is an ever present idea or fantasy whose absence is mourned and yet whose reality is never fulfilled.

Higgins describes infertility in terms of object relationships and its effects on individuals, couples, social networks, relatives and other social interactions. Olshansky theorizes about identity formation in infertile individuals in which infertility becomes a central focus of their self-definition, rather than other concepts such as career and relationships.

Many authors on this subject refer to the developmental model of Erickson in which the major task of middle age is generativity, "establishing and guiding the next generation". While parenthood is the major means of achieving this goal, there are other paths to such goals for both men and women, such as teaching children, having relationships with young people, mentoring younger colleagues at work.

A major pioneer in this field has been B. E. Menning who encouraged the idea that for many individuals and couples a grief reaction was a major response and that its resolution was the way that many could cope the best. Many writers have elaborated on these responses, and contended that "infertility is a chronic sorrow (for many) in which the pain of the loss is not forgotten, but periodically remembered and mourned again" (Burns, 1993).

For those individuals or couples who long for children, the commonest responses are anger, guilt and blame, sadness, depressive symptoms, and at times, loss of self-esteem and control over one's life. Many interactions with others, including spouse, family members, friends and co-workers, may be altered. Depressive syndromes may occur or obsessive-compulsive symptoms in which ideas about becoming pregnant and pregnancy inducing treatments become the center focus of one's life and thought. Marital/partner interactions may be disrupted especially if poor communication is part of the relationship. Sexual difficulties are especially common in infertile couples where treatment has been intrusive into the ordinarily private aspects of life. The most common are a diminution of sexual desire in both sexes. In women, dyspareunia may result from a lack of sexual arousal and lubrication, orgasmic response may diminish or be absent, and avoidance may occur. Men often experience erectile difficulties. The sexual relationship may be affected by such beliefs as that sexual problems are part of infertility, sacrificing the pleasure of sex may be punishment necessary to have a child (bargaining), and infertility resulted from past sexual behaviors (Burns, 1993). There is the hope that with the success of parenthood, sexual problems will be alleviated. Again for couples who can communicate better, these problems may be lessened.

Many couples do cope well and comply with many difficult and rigorous treatment regimes in their quest for a child. Hormones that are used to enhance ovulation may cause emotional lows and highs. The appearance of each menstrual cycle is a reminder that the body did not function in a way that was desired.

The psychological effects of infertility are shaped by a number of factors. These are related to one's age when infertility is diagnosed, stage of development, basic personality structure, coping style and defense mechanisms, pre-existing psychopathology, culture and religion, whether one is married or single, environmental supports, medical causes for the infertility, motivations for pregnancy, and very importantly, the skill of the medical staff who care for the individual, present her with information and options, do the procedures, and treatments. There is not any evidence from numerous studies to suggest that infertility patients have any increased occurrence of psychopathology more than the population in general, or have any particular personality style or disorder (Downey, 1989). Many cope extremely well with difficult situations, and are, in fact, more compliant and agreeable than might be expected.

An understanding of the psychology of infertility also requires that one be familiar with ways emotional conflicts can affect fertility, as well as how infertility affects psychological functioning.

For many years, psychogenic infertility was equated with unexplained infertility. They are not the same. One must think about this in historic context. As recently as the mid-1960's, only about 50% of infertility was known to have an organic cause. Today there are physical problems identified in about 85 to 90% or more of cases (Speroff, 1989). In a review of the earlier literature, psychogenic infertility was a way of pejoratively labeling women in whom no physical cause was found. These women were called unfeminine, rejecting of children and sexuality, or unmotherly. Their stress was multiplied many times as a result of these attacks on their identity. They were advised to get psychological treatment for their infertility. Today the emphasis has been on noting the stress that infertility has caused in the individual's life, as well as defining emotional problems (Menning, 1980).

There are psychological causes of infertility, but they are not diagnosed by an absence of a physical problem, and may co-exist with such disorders. Stress, which is often hard to define, may cause changes in reproductive function via the hypothalamic-pituitary-gonadal axis. An example might be anorexia nervosa or major depression with weight loss leading to amenorrhea and anovulation. Another example might be stress or bereavement leading to immune function alteration as illustrated in the work of Kiecolt-Glaser where medical students undergoing stress examination were noted to have a decrease in natural killer cell cytotoxicity correlating with perceived stressful life events and loneliness. Depressive symptoms, inadequate coping styles and minimal or restricted use of social supports has been associated with reduced measures of immune function. Sexual dysfunction can lead to infertility where disorders such as inhibited sexual desire, pain with intercourse, or vaginismus leads to a decrease in frequency or cessation of sexual activity. Erectile difficulties are frequent in men during infertility workup and treatment procedures. Much of the spontaneity of sexual activity is lost with couple's experiencing embarrassment, and self-consciousness at thinking about directions given to them regarding intercourse during such private moments.

The kind of diagnostic procedures couples go through may be experienced as intrusive with grading of very private aspects of one's life. The gynecologist takes a very detailed medical and sexual history from each partner. There are numerous tests, such as assessing the cervical mucus shortly after the couple have had coitus. These procedures and the treatments need not be dehumanizing if they are done in a sensitive and caring manner. The criticism of some of the procedures have more to do with how the staff provide them, rather than the procedures themselves. Ovulation inducing hormones may cause mood swings and emotional lability. Patients need to know that those are not unusual side effects. The response to treatments both physically and psychologically, vary with the same factors as those mentioned before, such as cause of the infertility, the type of procedures done, how the care is given, how long treatments must continue, success and failure. Another major problem may be repeated fetal loss.

Grief reactions may follow failed reproductive technology treatments such as *in vitro* fertilization where the woman thought of herself as pregnant. (Black, 1993; Greenfeld, 1988)

Secondary infertility, where a woman has had one or more children and is unable to conceive, can also bring with it much pain.

A. Assisted Reproductive Technology

With the birth of the first *in vitro* fertilization baby in 1978 in England, a new era began in reproductive biology and remarkable technology. While it brought great hope and promise, it has also meant disappointment and failure to many individuals. Those using these technologies have been very optimistic and believe they will be the ones to beat the odds and achieve a pregnancy. They present themselves in a very positive light no matter how badly they are feeling for fear of losing their last chances to try to achieve a pregnancy.

B. Roles for Mental Health Professionals

While most patients attending infertility clinics believe that there should be the provision of psychological services, a smaller number say that they would use these services, and an even smaller number actually utilize such programs. The reasons for this under utilization remain unclear, but probably relate to the stigma still attached to psychological treatments. Such services may include individual therapy, group therapy, self-help groups (such as Resolve), cognitive behavioral therapy or behavior therapies such as relaxation and self-hypnosis. Any therapist working with infertility patients should have a good background and knowledge about the subject, including acquaintance with the medical procedures and the psychological reactions. He or she should be able to talk with the gynecologists and the staff of the clinics, and be able to answer their questions as well as being able to ask appropriate questions.

The therapist should be able to explore the meaning of a pregnancy to a woman. Many conscious and unconscious conflicts may be involved and inability to conceive may be seen as a failure to achieve adult status, a narcisstic injury, a defect in one's body image or a loss of the ability to accomplish a developmental task in the life cycle. Infertility in a young, single person may indeed change the course of their life and relationships and intervention may be extremely helpful.

The goals of therapy are to help restore the sense of body wholeness and completeness, lessen the sense of narcisstic injury so as to lessen lasting bitterness (Kraft). This means working through the grief and loss, and understanding that the inability to conceive and carry a pregnancy to term does not indicate a psychological deficit or personal failure. The clinician must help patients to assess their future role, understand the place of parenthood in their lives, help them decide whether or not to proceed with further technology, adopt a child, or live without children. It is important that there be some resolution of these conflicts especially if the couple or individual plan to adopt or to live with children. Lasting anger can be destructive.

The gynecologists and staff of infertility clinics usually are appreciative of psychological help with their patients. Sometimes, however, they fail to recognize problems and may not make appropriate referrals, until their patients are in considerable emotional distress. In part, this may be because infertility patients tend to be particularly compliant and non-disruptive. They want to look like "good" patients so that they will be chosen to receive the "gift" of a pregnancy and child. They are afraid that if they acknowledge psychological problems, they will be rejected or abandoned. The doctor and nurse may be seen as powerful parental figures. Other patients may be difficult and present to the medical staff as angry and demanding, overly passive, depressed and guilt ridden, hopeless or overly optimistic, presenting with somatic complaints, or exhibiting an exaggerated fear of any or all procedures. They may be overly dependent calling the staff frequently. Such patients can be helped with better communication, education, coordinating medical strategies and staff talking to one another about treatment plans (Applegarth, 1989), but above all, recognizing their emotional difficulties and addressing them.

Men and women tend to express their emotions differently, with men typically saying less and being more self-contained. Women may be more open about their dysphoric feelings, and angry at the men who don't seem to care enough about the dilemma in which the couple find themselves. It is helpful for a mental health professional to meet all couples at the start of their infertility workup, educate them about some of the emotional reactions and such gender differences. It is helpful for the therapist to be present in the infertility clinic so that an implied message is that the staff of the clinic is supportive of psychological treatment.

The goal of these first meetings, whether alone or as a group, is to lessen the stress and tensions resulting from the diagnosis of infertility and its workup. The therapist must make clear that the goal of this therapy is not designed to lead to a pregnancy, although that would be wonderful if it happened. A careful history is helpful in trying to understand how the patients have coped with stress and loss in the past. History taking includes development, education, religious beliefs, jobs, friendships, psychiatric problems and alcohol or drug abuse. Reproductive history covers the patient's past pregnancies, if any, first realization of infertility, what brought her to the physician, what tests and procedures have been done and what have been her reactions to them, in whom has she confided about her problems, and what sort of emotional supports are available?

While sexual problems have been identified as both cause and effect, there have been very few careful studies and relatively few patients have been identified as having difficulties. Fagan and his colleagues examined 46 couples in an IVF program, and found a 15.5% rate of sexual dysfunction in at least 1 partner. In men, premature ejaculation and inhibited male orgasm were diagnosed, while in women, vaginismus, inhibited sexual desire and inhibited orgasm were noted. The treatment of such sexual dysfunctions include the use of behavioral modalities such as sensate focus exercises along with psychotherapy where indicated.

Brief psychotherapy as defined by Masterson and Davanloo and others may be an alternative for some patients. An average of 4 to 6 sessions are held with focus on the acute distress stirred up by confrontation with the implications and consequences of the infertility. Issues of loss, may also be a theme.

Bresnick (1981) characterizes 3 groups in her work as a counselor with infertility patients. One group has psychiatric conditions prior to the current situation. The second group may have milder psychiatric problems, but they have markedly increased recently. The third group are emotionally very stable, but feel dysphoric and want some more effective means of coping. She tailors the psychological treatment to the needs of each group.

Some special issues may arise in psychological work with infertility patients. Their relationship with their infertility physician and his or her staff are frequent sources of tension and concern. Patients are especially vulnerable, and feel guilty, blameworthy and angry about their medical status. They carefully scrutinize the medical staff for real or imagined slights, or signs of criticism or condemnation. Since many of the clinics are very busy, minor details are often overlooked. A patient may feel angry that a phone call was not returned or a laboratory result not reported. This sort of experience may echo the critical or judgmental responses of other authority figures, such as parents, teachers and employers.

Treatment decisions may be a very suitable topic for discussion with the mental health professional. The kinds of decisions relate to reproductive choices, diagnostic and treatment procedures, which treatments to undertake, how many times to undergo specific treatments, whether or not to consider new technologies, and when to stop treatment all together. When is it time to consider alternatives, such as living without children, or adoption? Some patients have difficulties because many of these options are so new, and untested that they have an aura of unreality about them. Was it easier, prior to the advent of new technology? Do some patients feel coerced by the choices?

If patients are suffering from a major depressive disorder, then the dilemma arises about whether or not to treat her with an anti-depressant drug while she is trying to conceive. The use of medications might affect the fetus if conception does occur. Patients also fear that the use of such medications might disqualify them from taking part in the infertility treatments. Anti-depressants do have the potential, though very small, for causing birth defects.

In summary, therapy for the infertile woman may be individual, group or partner based. The goal is to treat depression, enhance self-esteem, decrease guilt and blame, promote optimism and reality testing, and help with problem-solving, allow catharsis and ventilation, decrease feelings of isolation and loneliness and clarify and praise where clearly indicated. Hopefully, despite the problems, all of this can lead to personal growth and useful choices.

BIBLIOGRAPHY

Abbey, A., Halman, L. and Andrews, F.: 1993, 'Psychological treatment and demographic predictors of the stress associated with infertility', *Fertil Steril* **57**, 122–128.

Baran A. and Pannor R.: 1989, *Lethal Secrets, The Shocking Consequences and Unsolved Problems of Artificial Insemination*, Warner Books, New York.

Berg, B.J. and Wilson, J.F.: 1991. 'Psychological functioning across stages of treatment for infertility', *J Behav Med* **14**, 11–26.

Berger, D.: 1977, 'The role of the psychiatrist in a reproductive biology clinic', *Fertil Steril* **28**, 141–145.

Brody, E.M.: 1987, 'Reproduction without sex, but with the doctor', *Law, Medicine & Health Care* **15**, 152–155.

Burns L.: 1993, 'An overview of the psychology of infertility', *Infertility and Reproductive Medicine Clinics of No America* **4**, 433–454.

Dennerstein, L. and Morse, C.: 1988, 'A review of psychological and social aspects of *in vitro* fertilization', *J Psychosom Obstet Gynecol* **9**, 159–170.

Domar, A., Seibel, M. and Benson, H.: 1990, 'The mind/body program for infertility: A new behavioral treatment approach for women with infertility', *Fertil Steril* **53**, 246–249.

Downey, J. et al.: 1989, 'Mood disorders, psychiatric symptoms, and distress in women presenting for infertility evaluation', *Fertil Steril* **52**, 425–432.

Erickson, E.: 1980, *Identity and the Life Cycle*, W.W. Norton, New York.

Fagan, P. et al.: 1985, 'Sexual functioning and psychologic evaluation of *in vitro* fertilization and embryo transfer', *Fertil Steril 43*, 48–53.
Freeman, E.W. et al.: 1985, 'Psychological evaluation and support in a program of *in vitro* fertilization and embryo transfer', *Fertil Steril* **43**, 48–53.
Greenfeld, G. et al.: 1985, 'The role of the social worker in the *in vitro* fertilization program', *Social Work in Health Care* **10**, 71–79.
Harrison, K.L., Callan, V.J. and Hennessey, J.F.: 1987, 'Stress and semen quality in an *in vitro* fertilization program', *Fertil Steril* **48**, 633–636.
Higgins, B.S.: 1990, 'Couple infertility: From the perspective of the close/relationship model', *Family Relations* **39**, 81.
Holder A.: 1993, 'Legal and ethical issues in assisted reproductive technology', *Infertility and Reproductive Medicine Clinics of No America* **4**, 597–614.
Hollender, M. and Ford, C.: 1990, *Dynamic Psychotherapy: An Introductory Approach*, Am Psychiat Press, Inc, Washington D.C.
Kraft, A.: 1980, 'The psychological dimensions of infertility', *Am J Orthopysciat* **30**, 618–628.
Lauritzen, P.: 1993, *Pursuing Parenthood. Ethical Issues in Assisted Reproduction*, Indiana University Press, Bloomington.
Leiblum, S., Kenmann, E. and Lande, M.K.: 1987, 'The psychological concomitants of *in vitro* fertilization', *J Psychosom Obstet Gynecol* **6**, 165–178.
Mahlstedt, P. et al.: 1987, 'Emotional factors in the *in vitro* fertilization and embryo transfer process', *J of in Vitro Fertil & Embryo Transfer* **4**, 232–236.
Mahowald, M.B.: 1993, *Women and Children in Health Care. An Equal Majority*, Oxford University Press, New York.
Menning, B.E.: 1980, 'The emotional needs of infertile couples', *Fertil Steril* **34**, 313.
Milden, R.: 1985, *Infertility and the New Reproductive Technologies: Psychodynamic Specualtions*, paper presented at the American Psychological Association, Los Angeles.
Morse, C.A. and Van hall, E.V.: 1987, 'Psychosocial aspects of infertility: a review of current concepts', *J Psychosom Obstet Gynecol* **6**, 157–164.
Noyes, R. and Chapnick, E.: 1964, 'Literature of psychology and infertility: A critical analysis', *Fertil Steril* **15**, 543
Olshansky, E.F.: 1987, 'Identity of self as infertile: an example of theory-generating research', *Advances in Nursing Science* **9**, 54.
Reading, A.: 1989, 'Decison making and *in vitro* fertilization: the influence of emotional state', *J Psychosom Obstet Gynecol* **10**, 107–112.
Rosenthal, M.B.: 1993, 'Psychiatric aspects of infertility and assisted reproductive technologies', *Infertility and Reproductive Medicine Clinics of No America* **4**, 471–481.
Rosenthal, M.B.: 1985, 'Grappling with the emotional aspects of infertility', *Contemp OB/GYN* **7**, 97–104.
Seibel, M.: 1988, 'A new era in reproductive technology', *N Engl J Med* **318**, 828–834.
Sokoloff, B.: 1987, 'Alternate methods of reproduction: Effects on the child', *Clin Peds* 11–17.
Speroff, L. (ed): 1989, *Clinical Gynecology Endocrinology and Fertility*, Williams & Wilkins, Baltimore, Maryland.

Stanton, A.L. and Dunkel-Schetter, L.: 1991, 'Psychological adjustment to infertility', in *Infertility: Perspective From Stress and Coping Research.* Plenum Press, New York, pp. 3–16.

Taymor, M.L.: 1990, *Infertility: A Clinician's Guide to Diagnosis and Treatment*, Plenum Medical Book Company, New York.

Roden, J. and Collins, A. (eds): 1991, *Women and New Reproductive Technologies: Medical, Psychosocial, Legal and Ethical Dilemmas*, Lawrence Erlbaum Assoc., New Jersey.

SECTION TWO

THEOLOGICAL REFLECTIONS

JAMES F. KEENAN, S.J.

MORAL HORIZONS IN HEALTH CARE: REPRODUCTIVE TECHNOLOGIES AND CATHOLIC IDENTITY

I. INTRODUCTION

In addressing the topic of moral dilemmas and the moral identity of Catholic health-care institutions, we must recognize the insistence of contemporary moral theologians to attend to the specific nature and historical context of ethical issues. This insistence is particularly important for the topic treated here, because for too many years, normative guidelines for Catholic health-care prescinded from actual situations and circumstances.

This tendency to resolve moral dilemmas with inadequate attention to the historical or concrete has been practiced for almost three centuries in the naive objectivism found in the "manuals" of moral reasoning. These manuals, which singularly expressed Catholic moral teachings, were based on the assumption that moral truth is best attained by trying to find a rule or law that at once resolves a particular question and yet can be applied across time, cultures and circumstances with equal cogency, coherence and urgency. These manuals (see Gallagher, 1990) sought to map out those laws, rules and principles that equally applied without adequate regard to history, culture and other "circumstances". Their objectivism was a particularly ossified form of the classicist approach to theology (on this and the historical approaches to theology, see Harvey, 1989, p. 487) which at other times in its history was not as hostile to the specific nature of the agent.

The manualist method is important because it is effectively the cause of considerable critique of the recent Catholic teaching on reproductive technologies, *Donum Vitae*. To understand the horizon that Catholic health-care faces regarding its identity and the use of these technologies, then, we must first understand that method used in the teaching because it is at odds with the legacy of Thomas Aquinas and considerably different from the method endorsed by contemporary theologians.

Second, we will briefly see just how the magisterial document itself engages two different methods, by defining the object of moral concern in two different ways. In the third section, we examine traditional though not "objectivistic" insights to forge some common ground from which to view the horizon. Finally, in the fourth section, we find other related insights that might further assist Catholic health-care facilities as they prudently determine policies and practices.

II. MANUALIST LOGIC

The method that the manuals engaged had three important features. First, the moral method (see Mahoney, 1987, pp. 175–223) disengaged any "subjective" elements, as objectivity became defined as the antithesis of subjectivity. Instead of arguing that objectivity is undermined by any arbitrary suspension of reasoning (see Fuchs, 1984, pp. 29–41; 1987, pp. 117–133), the manualists named the agent-as-subject as the threat to moral reasoning and removed almost all circumstances surrounding the agent. Only recently, such circumstances (see Keane, 1982) have re-entered into objective estimations of right conduct.

Curiously, this anti-subject insight in moral logic found in the manuals is similar to the contemporary ethical writings at the universities. Alasdair MacIntyre (1981) argues forcefully that the Enlightenment project tried to find transhistorical and transcultural solutions to moral problems that could always be applied equally precisely by denying reason any "subjective" or "cultural" inclinations. He adds that the enlightenment project was complemented by the reformation which saw in the post-lapsarian human a reason so wounded that it was incapable of exercising prudence. Firm, fixed universal laws would provide the needed guidelines that the fallible, self-interested agent could not.

Thus, despite the assertions that Roman Catholic theology is less pessimistic than Protestant theology about the effects of original sin, in practice both theologies did not trust the agent. While Protestantism argued that prudence was nothing more than self-interested reflection, by the time of Francis Suarez, Roman Catholicism (see Treloar, 1991) abandoned the call to acquire prudence and imposed an "objectivistic" structure for the ordinary Catholic. Prudence was no longer the virtue of the ordinary person seeking to do the right; prudence became "*the* virtue of the legislator" (Ulshafer, 1992, p. 11). Moreover, the legislator did not simply articulate and present the principles; by the time of

Suarez, the legislator (Jonsen and Toulmin, 1988) applied the principles as well. He legislated not simply the law, but his determinations as well. Thus, probably the greatest contradiction in the history of Western moral thought is found from the seventeenth to twentieth centuries when two traditions, Roman Catholicism and the Enlightenment, upheld human reason while denying the agent the individual responsibility to reason rightly.

This contradiction becomes, today, a tragic irony when some Roman Catholic moral theologians, like Kiely (1980), contend that because agents are so incapable of prudential reasoning they ought only to accept the objective moral instructions given them from the magisterium. Instead of asking whether the incapacity comes from the failure of moral theologians for centuries to assert the responsibility that all Christians have to become prudential masters or mistresses of their daily conduct, Kiely's position further arrests the moral development of Christians and declares that the ordinary Christian, who attempts to think otherwise, is probably dysfunctional.

Second, in this anti-subjective method, the manualists changed the way the primary scholastic concept for moral reasoning, the "object", is used. The object (Kopfensteiner, 1992; Mullady, 1986; Stanke, 1984) is that concept by which a moral theologian or bishop defined moral activity.

In Thomas Aquinas's writings, the primary object that explained moral activity was the proximate content of one's intention rather than the physical action. Unlike the manualists's use of the object, that is, as derived from an external object, Thomas never defined the object as an act in its physicality, but rather in its agency: the object was not derived primarily as some explanation for an external action, but rather from the material content of the intention.

Some may want to invoke the difference between *finis operis* and *finis operantis* and argue that the object is equally found in an external action and an intention, respectively. But this language, at home in the manualist tradition, is really not the language of the thirteenth century *Summa Theologiae*. Thomas rather made, as Lottin, Kluxen, Pinckaers and Riesenhuber rightly insist, the object as the primary concept of moral description and found its first locus for evaluation in the intention. Moreover, the object was not something abstracted or extracted from an external action. Rather, the object was the material content of the intention which as intentional did not have the narrowness of definition found in an external act. Thus, the object of the intention was measured by Thomas as being either virtuous or vicious.

When Thomas wrote, for instance, about the object of lust (*Summa Theologiae* II-II. 153. 2c and 3c; 154. 1c) he did not look to the acts of rape, seduction or adultery; nor did he say that the intention is that which assumes the object of the acts of rape, seduction or adultery. Rather, he described lust as an attitude of unrestrained indulgence in venereal pleasure. That is, one does not become lusty because one commits the acts of seduction or rape, nor does one become lusty by assuming into the intention the object of these external acts. Rather one becomes lusty when the content of one's attitude or intention is disordered. For Thomas, the intention precedes external actions and the object is derived first from the intention, not from the external action. In a word, lusty intentions lead to seductive activity, not vice versa.

Thomas's position represented a development in the evaluation of moral action. His predecessors derived moral description from external actions, but Thomas arguing for the virtues realized that agents are guided by what they intend and then seek actions that may realize that intention. Objects were found in our thoughts, not in physical actions. Thus, Thomas's concept of intention did not convey a necessarily privatistic notion. For Thomas, the concept of end which described the intention was completely compatible with the concept of object. This was, in part, the achievement of Thomas, (see also Keenan, 1992, pp. 38–91) to successfully wrestle intention from subjectivity by arguing that the intention has its own object prior to and able to inform the external action.

Yet, in the manualist tradition, the pre-*Summa Theologiae* use of object re-emerges. There the object is derived from the external act and thus contemporary examples of how the object functions often appear distinctly with a manualist emphasis. In the manualist tradition, the intention, then, became defined by what object of an external activity is assumed by the agent. This is a reversal of the achievement of Thomas.

As an extension of the manualist tradition, a contemporary explanation of object appears as follows. Injecting someone with morphine can be explained as alleviating pain, supporting drug addiction, or hastening someone's death. Each of these explanations is an object of the physical activity of injecting someone with morphine. The moral evaluation, then, depends on which object has been assumed into the intention. The object, then, is that which gives intelligibility or content to an external action and that, in turn, can be evaluated as right, wrong, or indifferent.

Sometimes, however, an action has only one object and, therefore, only one intention can explain a particular activity. For instance, the object of injecting an air bubble into someone's veins can only be described as killing. Certainly one may argue that one is trying to release another from pain, but one is doing that precisely by killing. Thus, injecting an air-bubble is not open to other "objects" or explanations, and therefore, anyone intending to inject an air bubble into a person's veins could not primarily be intending anything other than killing.

Two cases further highlight how the concept of object functions in manualist logic. At its discovery the use of the "pill" was considered a contraceptive device. Later it was recognized as a method of correcting an irregular menstrual cycle. By this recognition (see this discussion, in Valsecchi, 1968, pp. 9–71) the object of taking the pill was now not only contraceptive but also therapeutic. One's intention, then, further determined whether taking the pill was contraceptive or therapeutic activity. Yet, in that activity, prior to the intention, two objects had been named and one's intention could only be an assumption of one or the other object.

Likewise, contemporary theologians have tried to redefine the object of activity associated with condom use. The object of using a condom was long considered (Denzinger, 1976, 2795, cf. 3187–89; 3638–40; Ford, 1943, pp. 578–84; 1944, pp. 518–520) a form of Onanism. Onanism held that any failure in heterosexual relations to allow semen to be deposited in the vagina is always wrong. Today in the midst of the AIDS crisis, however, the condom has become the prophylactic (Keenan, 1989, p. 213; Tuohey, 1990), that is, the object of activity is not necessarily contraceptive and can be described instead as disease preventive. Again, only an actual intention can determine which object, contraceptive or prophylactic, is being aimed at.

Defining the object allowed manualists the ability to distinguish licit from illicit forms of activity prior to any consideration of the actual intention or circumstances regarding someone's conduct. In the manualist tradition, the first moral action, then, is not determining one's intention, but rather, determining what licit objects are available. If one can not determine a licit object for an activity, then in the manualist tradition no intention can justify one's conduct.

In the manualist tradition, naming the object established the parameters for describing any possible intentional activity. That is, in the manualist tradition for the last three hundred years, for some activities, one could only describe one's intention according to the determinations. Thus, the physical action of prompting seminal emission, whether for pleasure, therapy against a prostate condition, medical examination for sexual dysfunction or other illness, or use in reproductive technologies, has been described in the manualist tradition simply as masturbation, (see Carlson, 1989, pp. 529–30, 538) an illicit use of the sexual faculty. Regardless of what one may claim to be the "intention", if a licit object is not available to express the intention licitly, there can be no rightly intended moral activity in the manualist tradition.

Here then the third feature. Moral theologians and bishops had the tasks not only of articulating principles and precepts and applying them to cases, but also of determining the object. Since moral intending depended primarily on the liceity of the object, much depended on how the object was articulated. Moreover, its expression usually determined which principle would be invoked and applied to determine the activity's liceity.

Since prompting seminal emission even for medicinal purposes has been always considered masturbation, then the principles prohibiting illicit sexual pleasure and illicit use of the sexual faculty are implicitly evident in the object's expression. This same issue faced the subject of removing testes in treating prostate cancer. For a period, this activity (removal of testes) had as its object, sterilization. Since one's reproductive organs existed for procreation, one could not subordinate the good of the species for one's individual good. Thus all objects of sterilization have been declared illicit by this principle of the priority of the species's good. Fortunately, Pope Pius XII described the activity by another "object" (Harvey, 1989, pp. 30–31); this object, concerning the treatment of an organ that harms the whole health of the body, was viewed as licit by the principle of totality. However, only when Pope Pius XII did define a new object for the activity could those suffering from prostate cancer have their testes licitly removed. With less good fortune, to this day, the variety of reasons women may give to have a tubal ligation have been singularly reduced to the object of illicit direct sterilization.

The object, then, even when it is derived from external actions, is capable of being a fairly flexible device. That flexbility is dependent upon the degree of sensitivity and prudential insight enjoyed by those with the power to define it. How it is defined, then, is pivotal for determining which suitable principle a legislator (moral theologian or bishop) should invoke to determine its liceity. But its role is not simply reserved to determining the liceity of individual moral actions or practices.

The object is equally the controlling insight underlying many moral reasoning structures found in the manuals. For instance, it is the foundational, first condition of the principle of double effect, that is, the object of activity must be good or morally neutral (see Ghoos, 1951, p. 43; Keenan, 1988; Mangan, 1949, p. 57). Likewise, it was central (see Keenan, 1989, pp. 210–217) in distinguishing the licit mediated material cooperation by one agent from the willed, illicit object of activity by another. In terms of my own argument that the notion of object has been given a different source by the manualists, it is important to note that these principles were articulated not by Thomas as some have argued (Mangan, 1949), but as Ghoos (1951) convincingly has demonstrated by seventeenth century writers, that is, by the founders of manualism. The object as used in these principles, that is, as derived from external activity is not the notion of object found in the intention and derived by Thomas from the virtues or vices.

Thus the most important work of the moral theologians who wrote the manuals was always in determining the object. Undoubtedly, that determination was never taken from an ahistorical context (see Kopfensteiner, 1992), but once the object was taken from a concrete situation, the manualists gave it a universal, transhistorical, transcultural significance. They sought to make the determination, barring other factors, the singular object or explanation for human activity.

This became particularly the case in the classification of intrinsic evils. Dedek narrates well the evolution of that doctrine, but central to the classification is the awareness that intrinsic evil is the determination of "pre-intentional" activity. That is, as described here, an intrinsic evil is simply the assertion by theologians and bishops that no licit object has been found by them for an action. Barring a new object for the action, the action remains always and absolutely prohibited. Again, on an historical note, Dedek (1983) shows that the concept of intrinsic evil originates not in Thomas's writings but a century later in the writings of Thomas's greatest Dominican detractor, Durandus.

Rather than critique the definition of the object as derived from the external action, the critique and subsequent debate in moral theology these last twenty years has concerned the unique role that the object holds. Apparently one reason why the critics do not critique the definition of the object is because such important yet diverse theologians as Janssens (1972), McCormick (1984, pp. 166ff.) and May (1984, pp. 582, 592) define the object precisely as the manualists did. They do not seem to assume that it had a more primary use.

Almost every major revisionist article addressed a reexamination of the manualists's claims that an object, as "pre-intentional", could sufficiently define an activity as always prohibited. Fuchs (1983, pp. 115–152), for instance, shattered that claim in his investigation into whether a universal norm could be declared a priori an absolutely, objective norm. His attack (see also 1984, pp. 71–90) was paralleled by Janssens's insistence that the morality of an action can never be determined prior to its concrete situation. At most, the object functioned as an ontic, not a moral, indicator. Likewise, Knauer's famous article on double effect argued that the object of moral activity could never be determined without an end. He went so far as to contend that the liceity of an object was determined by its commensurability with an end. For him, the end replaced the object as the measure of moral activity. Despite the gross oversimplifications attributed to the revisionism of these twenty years (for instance, Kiely, 1985 argues that the major revisors's works are consequentialist), Connery recognized that the revisionism of the 1970s and 1980s was an attempt to reconfigure the moral method by, notably, removing the absolutely determinative function of the object. If the moral method has been in any way revised it has been precisely in the reconfiguration of the object in the place of moral reasoning. But in this debate all seem to assume the manualist understanding of the object, that is, that the object is derived from an external action rather than the stuff of the intention.

The revisors's work (see especially, Cahill, 1981; Vacek, 1985) has been supported by the important studies of the method of moral reasoning immediately prior to manualism, that is, the high casuistry of the 16th century. Besides Mahoney's important work, Jonsen and Toulmin demonstrate that the deductive method employed by the manualists was not the primary method of reasoning for their predecessors. The casuists two-fold interests in the use of analogy to compare cases and in the speculation of newer principles both to ease restrictions on such professions as banking and to respect local cultures distinguished them from the more narrow method engaged by their successors. Thus, like Thomas, their "object" (Keenan, 1993a) was not a physical external; unlike Thomas, their object was a practice or a policy, not a virtuous or vicious attitude.

In summary, then the anti-subject tendency of the manualist method became reinforced by their particular interpretation of the object that was thoroughly apersonal or inhuman. Moreover, since the defining of the object and the applying and determining of principles to situations was in the hands of the legislator it is no surprise, then, that the manualists's judgments were viewed as considerably strict and narrow and hardly inclusive of the diverse insights, intentions, and circumstances that surround ordinary people's lives. In a word, the judgments are anti-historical.

III. *DONUM VITAE* AND MANUALIST LOGIC

The recent magisterial teaching, *Donum Vitae*, imposes not only particular principles or precepts but also the specific application of these principles to a variety of objects pertaining to reproductive technology. It prohibits nearly every form of major reproductive technology. Its best arguments are not against particular forms of technology, nor against particular actions, but rather against the attempts of persons who seek parenthood by the use of gametes from someone other than a spouse. The object here, of attempting to derive parenthood outside of the marital context, is, interestingly, derived from an intention, not an action. This object of heterologous fertilization and insemination contradicts several principles and, as the document makes clear, is "contrary to the unity of marriage, to the dignity of the spouses, to the vocation proper to parents, and to the child's right to be conceived and brought into the world in and from marriage (p. 24)."

The section that prompted the greatest controversy concerned the context where the gametes of married partners are the gametes of the parents. The Vatican prohibited reproductive technologies if they replace the marital act as the procreative act. As a result, the Vatican only permitted certain *methods* of artificial insemination by husband. Other methods of AIH as well as every type of *in vitro* fertilization in this homologous context are prohibited. Both of these prohibitions were based on defining an object derived from a physical type of action. For instance, regarding homologous *in vitro* fertilization, the document states, "The process of IVF and ET must be judged in itself" (p. 29) and "the very nature of homologous IVF and ET also must be taken into account" (p. 30). Because of this process, "the generation of the human person is objectively deprived of its proper perfection: namely, that of being the result and fruit of a conjugal act" (p. 30). The object is derived not from the intention of married partners but from the physical fact that it does not occur within a sexual act. That is, the object is derived from a particular type of external act. Thus, AIH is permitted if the type of technology used is doable within the context of a sexual act, that is, so long as the "technical means is not a substitute for the conjugal act but serves to facilitate and to help that the act attains its natural purpose" (p. 31). Its liceity depends, then, upon an object derived from an external action.

In *Donum Vitae*, the action of these reproductive technologies has been completely divorced from the homologous context, that is, precisely the fact that the intended genetic parents are married is specifically omitted. Subjectivity is not being omitted here; what is specifically denied from the determination of this activity's liceity is the objective fact that husband and wife are seeking to become father and mother. That fact is as "objective" as the fact that only a husband and wife engage in conjugal activity. That is, in the Catholic tradition being husband and wife is neither a subjective nor a circumstantial condition but rather the specific objective fact that determines permitted sexual relations. The liceity of their activity absolutely, that is, objectively depends on whether the participants are married. That insight precedes the extraordinarily "objectivistic" interpretation of the tradition now being exercised by the manualist mentality. In *Donum Vitae*, the act itself appears to propose explanation. This is particularly problematic when the very definition of a "conjugal" act is derived from the agents themselves.

The document employs, then, two methods. One enjoys considerable consensus as it argues that (biological) parenting cannot be intentionally separated from marriage. The other derives its argument from a physical action and has received considerable critique. I want to conclude predicting that to the extent that the horizon of Catholic health-care is proscribed by the second method, to that extent the horizon will not be an illuminating one, but a divisive one.

IV. SCHIZOPHRENIC HORIZONS?

In light of these investigations, then, the moral horizons in Catholic health-care as it faces reproductive technology are in a nearly schizophrenic context. Though Cahill points out often (e.g., 1989) that the disagreements between methods need not be considered an impasse, nonetheless, the intensity of the critique of the document consistently focuses on the logic of excluding homologous matters. A variety of critiques (Harvey, 1989; McCormick, 1987; Trau, 1990b; Vacek, 1988) attempt to persuade the Vatican to maintain its ban against heterologous matters, but to permit homologous fertilizations especially, if the problem of "spares" is overcome. Thus, the tension is not between theologians and the magisterium, the tension is between two different methods used by the magisterium.

Despite these difficulties, it is helpful to realize that the horizons of secular health care is devastatingly worse, being bereft of any material principles, save perhaps particular expressions of justice. As MacIntyre notes, the enlightenment project failed and in the absence of universal principles the principle of autonomy continues to emerge as the singular controlling principle in medical ethics.

Moreover, whereas the concept of object is unnecessarily restrictive in the manualist tradition and in fact helps support the greatest aberration within the tradition, that is, the attempt to make the locus for measuring moral activity outside of an agent's intention, still the document *Donum Vitae* avoids these difficulties and enables both members of the Church and of society at large, precisely when it engages subjects and when it derives objects from intentions rather than external actions.

Thus the document rightly argues that an attitude of a "right to a child" is wrong, but the document overstates itself when it implies that the action of *in vitro* fertilization (in any context) makes the child an object (p. 34). Likewise the document rightly warns against an intention of "domination" (pp. 5–6, 22, 30), but cannot prove that *in vitro* fertilization of a married couple "establishes the domination of technology over the origin and destiny of the person" (p. 30). These incongruencies are not only evident in the document; they are at the source of almost every magisterial conflict within the Catholic health-care community.

When the magisterium argues that natural family planning makes birth control licit while another activity does not, it engages the object as derived from an external action. The object of the external action of natural family planning does not involve separating the unitive and procreative ends of marriage. That reasoning, which only emerged in the manualist tradition and in some of the pre-*Summa Theologiae* writings, stands as the weakest expression of Roman Catholic moral logic. Far more compelling is the magisterium and the tradition when it prohibits (as it always has) the intention to exclude children from marriage. Likewise, when the magisterium argues that cutting a fallopian tube justifies the removal of an ectopic pregnancy, it derives the object from an external action and grounds moral rightness in an act. Far more credible was the magisterium, when it tolerated only abortions based on a woman's justification to save her own life against a life-threatening pregnancy. But the questionable teachings, like that which finds the moral liceity of AIH when a perforated condom is used, are notably recent ones; they reflect the three factors of the manualist tradition named earlier. As a result, the magisterium must attend not simply to these particular judgments, but much more importantly to certain expressions of moral logic that undermine the important credibility it has to teach moral principles. The magisterium needs to familiarize itself with a logic within the moral tradition much more cogent than the manualist method. Thomas, the casuists, the patristics, and most importantly, the *Gospels* taught a moral logic based on what we intend; they did not teach a moral logic based on physical or external actions. Their teachings illuminated the horizon of moral insight.

V. TRADITIONAL SUPPORTS FOR THE FUTURE

Aside from those exceptions wherein the object of moral evaluation is derived from external actions, the moral logic of the Church's teaching tradition is extraordinarily healthy and strong. It demands that Catholic health-care centers must intend to serve and treat all whom the centers can; it treats its employees with as much respect and due privileges as justice demands; it prohibits intending the death of any patient, whether a fetus or an autonomous adult requesting assisted suicide; it argues that only those who are married can properly intend children; it refuses to demand the use of extraordinary methods of life support. The horizon for Catholic health-care only suffers *in its identity* when it is divided by teachings employing a more problematic method of proceeding.

Still three topics merit our attention for facing the future. First, especially problematic issues, e.g., reproductive technologies, the use of fetal tissue and genetics, will require that Catholic physicians, researchers, nurses, theologians, and others to be informed. A major set-back in this area has been due to the removal of Catholic hospitals from research and practice particularly in reproductive technology. Beyond the issue of the immoral use of spares, questions about either the right use of resources or psychological health (Fagan, 1986) may lead many to doubt the moral validity of *in vitro* fertilization in a marriage. On the other hand, the treatment of spares and now the practice of "embryonic reduction" by non-Catholic facilities suggests that the absence of Catholic medical competence is noteworthy. Were Catholic hospitals involved in reproductive technologies could they not pursue policies that would not only reject the practice of embryonic reduction but also do away with any necessity for spares? In a word, intolerance regarding Catholic research facilities in *in vitro* fertilization has meant that Catholic thinkers are not interlocutors with the researchers.

It would not be appropriate, however, to describe the Vatican as only intolerant regarding such research. Though the only American Catholic facility conducting *in vitro* fertilization ceased its work because of *Donum Vitae*, a number of European Catholic health-care facilities did not. Their continued research and practices suggests at least a tolerant Vatican. In a similar vein, it does not seem that the at times intolerant and punitive stance that the Vatican took toward a variety of theologians in the early eighties is as evident in light of the continued criticism not of *Donum Vitae* per se, but of its specific argument on AIH and homologous fertilization. This could be another expression of tolerance.

Toleration ought not to be confused with condescending patience. The tolerant person recognizes on the one hand that her/his judgment protects particular values that are presently threatened but that the judgment is at odds with other's. Yet, in the interest of moral objectivity, toleration (see Demmer, 1982; Post, 1968) recognizes the somewhat provisional nature of moral judgments. Moreover, it seeks to avoid unnecessary suppressions and rather seeks to allow the data of human experience to express itself through a sharing of opinions. By the same token, it does not relinquish its obligation to protect certain moral values and abiding insights. Thus differently from other attitudes, tolerance is not a form of patience or of disapproval, but rather an admission that in the present limited situation the obligation to protect certain values overrides the ability to recognize as right or acceptable other's decisions in conscience to live or act as they do. Thus the tolerant attitude is one of hope that one day, greater understanding will resolve the present incompatibility or threat and lead somehow to the reconciliation of which Paul calls us all to be ministers (2 Cor 5:11–6:13). In the meanwhile, it acknowledges that another's judgment is at odds with its own but does not suppress the other. In light of trying to articulate right and wrong intentions regarding reproductive technologies, the principle of tolerance provides, then, a traditional support.

Second, in this same vein, the interests of economics and justice will prompt greater occasions for joint ventures among health-care facilities. In these contexts, directors of Catholic health-care facilities will find themselves often in ventures in which the attitudes of the directors of other facilities will not simply be incompatible with but will actually contradict their own attitudes. Catholic attitudes on respect for life, welcoming the indigent, and promoting parenthood are particularly unique when faced with certain aspects of American health-care that give priority to the principle of autonomy. For this reason, the principle of cooperation (the conditions can be found in: Davis, 1958, pp. 342–352; Keenan, 1989; Kelly, 1958, pp. 332–335) could provide considerable insight for Catholic health-care facilities as they seek to balance their roles in joint ventures. The principle helps distinguish between what justice demands and what scandal prohibits. The principle does not necessarily permit or prohibit; rather it guides and provides its users (for example, see Griffin, 1990; Maestri, 1992; Ulsfafer, 1992) insight into the importance of demarcating or line drawing.

Cooperation does not necessarily have to hold the manualist reading of the object. Unlike the principle of double effect, the object in the principle of cooperation can concern intentions, rather than external actions. The principle of double effect (Keenan, 1993b), a terribly overused guide of extraordinarily dubious merit, is burdened by the fact that the double effect is derived from an external action and not an intention. The locus is precisely on the object as defined by the manualists, who (as noted above) articulated the principle in the first place. Outside of that logic, then, the principle is irredeemable.

Cooperation provides richer assistance if we are not bound to the narrow meaning that arise from actions. If cooperation uses the manualist notion of object, then we will be drawing lines about what actions we perform, instead of what values we want to see protected, promoted, and expressed. If the realm of cooperation does not treat the values we and the others have in intention, we may find ourselves cooperating blindly. Placing the object of cooperation in the realm of intentionality provides a context for honest discussion and disagreement and allows us to know better to what extent we should and should not enter into joint ventures.

Specifically in the area of reproductive technology, questions regarding attitudes toward unborn human life and the unique call to parenthood in the context of marriage must be addressed. Cooperation does not require that the same attitudes must be shared. On the contrary, cooperation is invoked precisely because they are not, but the degree of the differences should help to determine, in part, the degree of the cooperation.

Finally, while appropriating from the Catholic tradition right attitudes, intentions, and principles, Catholic health-care facilities need to get into the practice of exercising the virtue of prudence. This is not the virtue of caution, nor is it the simple application of principles to situations. That interpretation again is derived from the manualist tradition. Rather prudence (Nelson, 1992) is the virtue of right reason and vision that requires setting long and short term goals that concretely express the values that particular Catholic health-care facilities seek. The leaders of Catholic health-care facilities must begin to articulate their expectations. Why will their costs and expenses be allocated for some research and/or services as opposed to others? How will the goals of a facility reflect genuinely its catholic identity? Are the goals sufficiently expressed so as to provide guidance in determining the degree of cooperation in joint ventures? Internally, does the facility provide enough moral instruction to its physicians, nurses, staff and researchers so as to enable them to make prudential judgments as well as set prudential goals?

Specifically in the realm of reproductive technologies, prudence demands that policies for these programs be determined in the context of the particular facility's mission. That mission, a prudential articulation of the facility's goals, ought to determine whether and to what extent the particular facility extends its resources into perfecting the technologies that in turn perfect the possibilities for marital love to become parental love.

As Catholic health-care facilities look to the horizon they will see a challenge to be competent in a number of areas. But central will be an ability to communicate within its own walls, within the Church, and within society at large its own vision, its own reason for being there. That articulation is not known by simply writing a charter; it is expressed by prudent trustees, physicians, CEOs, and nurses who know why they belong to a facility that professes to be Catholic. Thus, the primary task for Catholic health-care facilities as they face a future maintaining moral and medical competency and Catholic identity should be to establish programs through which their members acquire and exercise the virtue of prudence. It will have to be the virtue of its members, after all, that guides the Catholic health-care facilities to right conduct.

BIBLIOGRAPHY

Bernardin, J.: 1987, 'Science and the creation of life', *Origins* **17**, 21, 23–26.
Byk, C.: 1989, '*Donum Vitae*: Civil laws and moral values', *The Journal of Medicine and Philosophy* **14**, 561–574.
Cahill, L.S.: 1989, 'Moral traditions, ethical language, and reproductive technologies', *The Journal of Medicine and Philosophy* **14**, 497–522.
Cahill, L.S.: 1981, 'Teleology, utilitarianism, and Christian ethics', *Theological Studies* **42**, 601–629.
Canadian Bishops: 1991, 'Reproductive technologies and the value of human life', *Catholic International* **2**, 633–637.
Carlson, J.: 1989, '*Donum Vitae* on homologous interventions: Is IVF-ET a less acceptable gift than 'gift'?', *The Journal of Medicine and Philosophy* **14**, 523–540.
Congregation for the Doctrine of the Faith: 1987, *Instruction on Respect for Human Life in Its Origin and on the Dignity of Procreation*, Libreria Editrice Vaticana, Vatican City.
Connery, J.: 1981, 'Catholic ethics: Has the norm for rule-making changed?', *Theological Studies* **42**, 232–250.
Connery, J.: 1983, 'The teleology of proportionate reason', *Theological Studies* **44**, 489–496.
Davis, H.: 1958, *Moral and Pastoral Theology*, Sheed and Ward, London.
Dedek, J.: 1977 'Moral absolutes in the predecessors of St. Thomas', *Theological Studies* **38**, 654–680.
Dedek, J.: 1979, 'Intrinsically evil acts: An historical study of the mind of St. Thomas', *The Thomist* **43**, 385–413.
Dedek, J.: 1983, 'Intrinsically evil acts: The emergence of a doctrine', *Recherches de theologie ancienne et medievale* **50**, 191–226.
Demmer, K.: 1982, 'Der anspruch der toleranz', *Gregorianum* **63**, 701–720.
Denzinger-Schoenmetzer: 1976, *Enchridion Symbolorum Definitionum et Declarationum de Rebus Fidei et Morum*, 36th ed., Herder, Rome.

Fagan, P. et al.: 1986, 'Sexual functioning and psychologic evaluation of *in vitro* fertilization couples', *Fertility and Sterility* **46**, 668–672.
Ford, J.: 1943 'Notes on moral theology', *Theological Studies* **4**, 578–84.
Ford, J.: 1944 'Notes on Moral Theology', *Theological Studies* **5**, 518–20.
Fuchs, J.: 1983, *Personal Responsibility and Christian Morality*, Georgetown University Press, Washington, D.C.
Fuchs, J.: 1984, *Christian Ethics in a Secular Arena*, Georgetown University Press, Washington, D.C.
Fuchs, J.: 1987, *Christian Morality: The Word Becomes Flesh*, Georgetown University Press, Washington, D.C.
Gallagher, J.A.: 1990, *Time Past, Time Future: An Historical Study of Catholic Moral Theology*, Paulist Press, Mahwah, New Jersey.
Ghoos, J.: 1951, 'L'acte a double effet, etude de theologie positive', *Ephemerides Theologicae Lovanienses* **27**, 30–52.
Griffin, L.: 1990, 'The church, morality and public policy', in C. Curran (ed.), *Moral Theology: Challenges for the Future*, Paulist Press, New York, pp. 334–339.
Harvey, J.C.: 1989a, 'Speculations regarding the history of *Donum Vitae*', *The Journal of Medicine and Philosophy* **14**, 481–492.
Harvey, J.C.: 1989b, 'A doctor reflects upon donum vitae', *Catholic Medical Quarterly*, 25–32.
Jabbari, D.: 1990, 'The role of law in reproductive medicine', *Journal of Medical Ethics* **16**, 35–40.
Janssens, L.: 1972, 'Ontic good and moral evil', *Louvain Studies* **4**, 115–56.
Janssens, L.: 1982, 'St. Thomas and the question of proportionality', *Louvain Studies* **9**, 26–46.
Janssens, L.: 1987, 'Ontic good and evil', *Louvain Studies* **12**, 62–82.
Jonsen, A. and Toulmin, S.: 1988, *The Abuse of Casuistry*, University of California Press, Berkeley.
Keane, P.: 1982, 'The objective moral order: Reflections on recent research', *Theological Studies* **43**, 260–278.
Keenan, J.: 1993a, 'The casuistry of John Major, nominalist professor of Paris (1506–1531)', *The Annual of the Society of Christian Ethics*, Georgetown University Press, Washington, D.C.
Keenan, J.: 1993b, 'The function of the principle of double effect', *Theological Studies* **54**, 288–313.
Keenan, J.: 1992, *Goodness and Rightness in Thomas Aquinas's Summa Theologiae*, Georgetown University Press, Washington, D.C.
Keenan, J.: 1989, 'Prophylactics, toleration, and cooperation: Contemporary problems and traditional principles', *International Philosophical Quarterly* **29**, 205–220.
Keenan, J.: 1988, 'Taking aim at the principle of double effect', *International Philosophical Quarterly* **28**, 201–205.
Kelly, G.: 1958, *Medico-Moral Problems*, Catholic Hospital Association, St. Louis.
Kiely, B.: 1985, 'The impracticality of proportionalism', *Gregorianum* **66**, 655–686.
Kiely, B.: 1980, *Psychology and Moral Theology*, Gregorian University Press, Rome.
Kluxen, W.: 1980, *Philosophische Ethik bei Thomas von Aquin*, Felix Meiner, Verlag, Hamburg.

Knauer, P.: 1967, 'Das rechtverstandene prinzip von der doppelwirkung als grundnorm jeder gewissensentscheidung', *Theologie und Glaube* **57**, 107–33.

Kopfensteiner, T.: 1992, 'Historical epistemology and moral progress', *The Heythrop Journal 33*, 45–60.

Kymlicka, W.: 1993, 'Moral philosophy and public policy: The case of NRTs', *Bioethics* 7, 1–26.

Lottin, O.: 1921–22, 1923 'Les elements de la moralite des actes chez Saint Thomas D'Aquin', *Revue neo-scolastique* **23–24**, 281–313, 389–429; 25, 20–56.

MacIntyre, A.: 1981, *After Virtue: A Study in Moral Theory*, University of Notre Dame Press, Notre Dame, Indiana.

Maestri, W.: 1992, 'Abortion in Louisiana, Act II: Prudence over passion', *The Linacre Quarterly* **59.2**, 37–43.

Mahoney, J.: 1987, *The Making of Moral Theology: A Study of the Roman Catholic Tradition*, Clarendon Press, Oxford.

Mangan, J.: 1949, 'An historical analysis of the principle of double effect', *Theological Studies* **10**, 41–61.

May, W.: 1984, 'Aquinas and Janssens on the moral meaning of human acts', *The Thomist* **48**, 566–606.

May, W.: 1988, 'The simple case of *in vitro* fertilization and embryo transfer', *The Linacre Quarterly* **55**, 29–36.

McCormick, R.: 1984, *Notes in Moral Theology, 1980 through 1984*, University Press of America, Lanham, Maryland.

McCormick, R.: 1987, 'The Vatican document on Bioethics: Some unsolicited suggestions', *America* **156**, 24–28.

Michael, M.: 1990, 'Screening for genetic disorders: Therapeutic abortion and IVF', *Journal of Medical Ethics* **16**, 43–47.

Mullady, B.: 1986, *The Meaning of the Term "Moral" in St. Thomas Aquinas*, Libreria Editrice Vaticana, Vatican City.

Nelson, D.: 1992, *The Priority of Prudence*, Penn State Press, University Park, Pennsylvania.

Pinckaers, S.: 1981, 'Le role de la fin dans l'action morale selon Saint Thomas', in *Le Renoveau de la Morale*, Casterman, Tournai, pp. 114–143.

Post, W.: 1968, 'Tolerance', in *Sacramentum Mundi VI*, Burns and Oates, London, pp. 262–267.

Riesenhuber, K.: 1971, *Die Transzendenz der Freiheit zum Guten*, Berchmanskolleg Verlag, Munich.

Riga, P.: 1987, 'The Vatican's instruction on human life', *The Linacre Quarterly* **54**, 16–21.

Stanke, G.: 1984, *Die Lehre von den 'Quellen der Moralitaet'*, Friedrich Pustet, Regensburg, Germany.

Trau, J.: 1990 'God-Talk, physicalism, and technology: A mutual endeavor', *Research in Philosophy and Technology* **10**, 291–295.

Trau, J.: 1990, '*Humanae vitae* and the current instruction on the origins of human life', *Research in Philosophy and Technology* **10**, 233–242.

Treloar, J.: 1991, 'Moral virtue and the demise of prudence in the thought of Francis Suarez', *American Catholic Philosophical Quarterly* **65**, 387–405.

Tuohey, J.: 1990, 'Methodology or ideology: The condom and a consistent sexual ethic', *Louvain Studies* **15**, 53–69.
Ulshafer, T.: 1992, 'The morality of legislative compromise', *The Linacre Quarterly* **59.2**, 10–26.
Vacek, E.: 1985, 'Proportionalism: One view of the debate', *Theological Studies* **46**, 286–314.
Vacek, E.: 1988, 'Vatican instruction on reproductive technology', *Theological Studies* **49**, 110–131.
Valsecchi, A.: 1968, *Controversy: The Birth Control Debate 1958–1968*, Geoffrey Chapman LTD., Washington, D.C.

WILLIAM E. MAY

DONUM VITAE: CATHOLIC TEACHING CONCERNING HOMOLOGOUS *IN VITRO* FERTILIZATION

I. INTRODUCTION

July 25, 1978 is a memorable date. First of all, it marked the tenth anniversary of Pope Paul VI's encyclical on marriage, *Humanae Vitae*, in which he affirmed that "there is an unbreakable connection [*nexu indissolubili*] between the unitive meaning and the procreative meaning [of the conjugal act], and both are inherent in the conjugal act. This connection was established by God, and Man is not permitted to break it through his own initiative" (Paul VI, 1968, n. 12). The significance of Pope Paul's claim for assessing the morality of homologous *in vitro* fertilization will be a matter of central concern in this paper. This date is further notable in that it is the birthday of Louise Brown, the "test tube" baby, the miracle child of modern reproductive technology. Louise was the first baby to be born after having been conceived outside her mother's body. Her birth, and indeed her conception, would not have been possible had the technology separating "baby making" from "love making" – the procreative and unitive meanings of the conjugal act – not been developed.

Pope Paul's concern in *Humanae Vitae* was with contraception and not with the laboratory generation of human life. Yet his teaching on the "unbreakable connection" between the two meanings of the conjugal act plays a central role, as will be seen, in the 1987 *Instruction on Respect for Human Life in Its Origin and on the Dignity of Procreation* – called *Donum Vitae* in Latin – in which the Congregation for the Doctrine of the Faith (hereafter CDF) formally addressed the moral issues raised by new reproductive technologies. This document, drawing on the understanding of marriage and human procreation found in the Catholic theological tradition, insists that the generation of human life, if it is to respect the dignity of both parents and children, "must be the fruit and sign of the mutual self-giving of the spouses, of their love and fidelity" (CDF, 1987, II, A, 1).

In its treatment of heterologous fertilization, in which gametes, whether ova or sperm, from persons other than the spouses are used to generate new human life, the *Instruction*, not surprisingly, concludes that this way of bringing new human life into being is gravely immoral. It is so because such fertilization is "contrary to the unity of marriage, to the dignity of the spouses, to the vocation proper to parents, and to the child's right to be conceived and brought into the world in and through marriage" (CDF, 1987, II, A, 2).

Although some find even this judgment of the *Instruction* too restrictive of human freedom,[1] many people, Catholic and non-Catholic as well, can appreciate the reasons behind it, even if, in some highly unique situations, they might be ready to justify heterologous modes of generating human life. Nonetheless, most people recognize that when a man and a woman marry they "give" themselves exclusively to each other and that the "selves" they give are sexual and procreative beings. Just as they violate their marital commitment by attempting, after marriage, to "give" themselves to another in sexual union, so too they dishonor their marital covenant by choosing to exercise their procreative powers with someone other than their spouse, the person to whom they have given themselves, including their power to procreate, "forswearing all others."

But many of these same people, Catholic as well as non-Catholic, find the *Instruction's* claim that even the "simple case" of *in vitro* fertilization and embryo transfer is morally wrong much too rigorous and harsh. In this case there is no use of gametic materials from third parties; the child conceived is genetically the child of husband and wife, who are and will remain its parents. In this case there need be no deliberate creation of "excess" human lives that will be discarded (perhaps through a procedure that some euphemistically call "pregnancy reduction"[2]), frozen, or made the objects of experimentations of no benefit to them. In this case there need be no intention to monitor the developing child *in utero* with a view toward its abortion should it develop some abnormality. Nor need there be, in this case, the use of masturbation – a means judged intrinsically immoral by the Catholic magisterium – in order to obtain the father's sperm, inasmuch as his sperm can be retrieved in nonmasturbatory ways. In this case, apparently, there is only the intent to use modern technology to help a married couple, incapable of having a child of their own because the wife's fallopian tubes are blocked or the husband's sperm production is low or because of other reasons, have a child of their own, to whom they can give the home where it can take root and grow under the loving tutelage of its own mother and father. Many people, including several Catholic theologians (e.g., Shannon and Cahill, 1988; McCormick, 1989; Verspieren, 1987), believe that recourse to *in vitro* fertilization and embryo transfer in this "simple case" is fully legitimate, since it does not seem to violate any one's rights but, to the contrary, seems to help a married couple's love blossom into new life. They quite reasonably ask what is morally offensive here? What evil is being willed and done? Is not the Church's magisterium being too harsh and rigoristic on this matter? Is it not insensitive to the agony experienced by involuntarily sterile married couples who are simply seeking to realize one of the goods of marriage by making intelligent use of modern technology?

In this paper I will first present the principles set forth in the Vatican *Instruction* to support its conclusions. Since the *Instruction* does not, in general, seek to establish the truth of these principles or show their special place within the Christian view of human life, I will then attempt to show their truth and reasonableness[3] and to respond to the major objections raised by Catholic theologians to the position taken by the *Instruction* on homologous *in vitro* fertilization and embryo transfer. In conclusion, I will seek to set forth what I consider to be the principal theological reason supporting the teaching of the *Instruction*.

II. THE REASONING OF THE INSTRUCTION

The *Instruction* presents three principal lines of reasoning to support its conclusion that married couples ought not resort to *in vitro* fertilization and embryo transfer, even if the wife provides the ovum and her husband provides, in a nonmasturbatory way, the sperm used to fertilize it.

The first line of reasoning appeals to the "inseparability principle" that, as we have seen, is at the heart of Pope Paul VI's argument against contraception in *Humanae Vitae*. Applying this teaching to the question of homologous artificial fertilization or the "simple case" with which we are concerned, the *Instruction* affirms, citing Pope Pius XII (1956, 470), that "it is never permitted to separate these different aspects to such a degree as positively to exclude either the procreative intention [as is done in contraception] or the conjugal relation" (CDF, 1987, II, B, 4, a). "Thus," the *Instruction* concludes,

fertilization is licitly sought when it is the result of a "conjugal act which is *per se* suitable for the generation of children to which marriage is ordered by its very nature and by which the spouses become one flesh" (*Code of Canon Law*, 1983, can. 1061). But from the moral point of view procreation is deprived of its proper perfection when it is not desired as the fruit of the conjugal act, that is to say, of the specific act of the spouses' union (CDF, 1987, II, B, 4, a).

According to this argument, it is morally wrong for married couples to generate human life outside the marital act, because by doing so they freely choose to sever the bond between the unitive and procreative meanings of the marital act. But the choice to do this deprives procreation of the goodness that it is meant to have as the fruit of the marital act.

A second argument presented in the *Instruction* to support its conclusion that homologous *in vitro* fertilization and embryo transfer is immoral is based on the dignity of the child so conceived. The Vatican document insists that the child "cannot be desired or conceived as the product of an intervention of medical or biological techniques" since "that would be the equivalent of reducing him to an object of scientific technology. No one may subject the coming of a child into the world to conditions of technical efficacy which are to be evaluated according to standards of control and dominion" (CDF, 1987, II, B, 4, c). But, the *Instruction* continues,

[C]onception *in vitro* is the result of the technical action which presides over fertilization. Such fertilization is neither in fact achieved nor positively willed as the expression and fruit of a specific act of the conjugal union. In homologous IVF and ET, therefore, even if it is considered in the context of *de facto* existing sexual relations, the generation of the human person is objectively deprived of its proper perfection, namely, that of being the result and fruit of a conjugal act in which the spouses can become "cooperators with God for giving life to a new person" (John Paul II, 1981, n. 14) (CDF, 1987, II, B, 5).

This argument can be summed up as follows: to desire or cause a child as a product of a technique is to make the child an object. But this is not compatible with the personal dignity of the child, who is just as equally a person as are his or her parents.

A third line of reasoning given in the *Instruction* to support its conclusions is based on the "language of the body." According to the *Instruction*,

spouses mutually express their personal love in the "language of the body," which clearly involves both "spousal meanings" and parental ones. The conjugal act by which the couple mutually express their self-gift at the same time expresses openness to the gift of life. It is an act that is inseparably corporal and spiritual. It is in their bodies that the spouses consummate their marriage and are able to become fathers and mothers (CDF, 1987, II, B, 4, b; cf. John Paul II, 1980, 148–152).

The document then concludes:

> In order to respect the language of their bodies and of their natural generosity, the conjugal union must take place with respect for its openness to procreation, and the procreation of a person must be the fruit and result of married love. The origin of the human being thus follows from a procreation that is "linked to the union, not only biological but also spiritual, of the parents, made one by the bond of marriage" (John Paul II, 1983, 393). Fertilization achieved outside the bodies of the couple remains by this very fact deprived of the meanings and values which are expressed in the language of the body and in the union of married persons (CDF, 1987, II, B, 4, b).

This argument holds that *in vitro* fertilization, which occurs outside the body of the mother and independently of the bodily act by which husband and wife embody their marital union in a unique and spousal way, is a way of generating human life that fails to respect the "language of the body," that refuses to acknowledge the deep human significance of the personal gift, bodily and spiritual in nature, of husband and wife to one another in the marital act.

The argument based on the "language of the body" – an argument rooted in the "theology of the body" set forth by Pope John Paul II[4] – is, in my opinion, intimately linked to the argument based on the "inseparability" principle. Thus in what follows I will seek to show how these two lines of reasoning seem to merge and how closely they depend on a basic vision of marriage, the marital act, and the generation of human life.

Of the three lines of reasoning found in the *Instruction* to support its conclusions regarding the absolute immorality of homologous (and, a fortiori, heterologous) *in vitro* fertilization and embryo transfer, I believe that the second, which rejects the generation of human life in this way on the grounds that bringing new human life into being in the laboratory is a form of production and depersonalizes human life by treating it as if it were a product, provides the basis for the most straightforward argument against resorting to the laboratory generation of human life. Nevertheless, I think that the other two lines of reasoning shed light on the wider issues regarding human existence raised by new reproductive technologies. But to understand the reasoning involved it is necessary to probe the meaning of marriage and the intimate links between marriage, the marital act, and the generation of human life. It is necessary to understand the fundamental vision of these realities that lies at the heart of the reasoning employed.

III. THE REASONABLENESS OF THE *INSTRUCTION'S* CLAIMS

A. *Marital Rights and Capabilities, the Marital Act, and the Generation of New Human Life*

"Not uniqueness establishes the marriage, but marriage establishes the uniqueness." These words of the German Protestant theologian Helmut Thielicke (1963, p. 95) are full of meaning. They express the truth that an individual man and woman, prior to marriage, are separate individuals, free to go their respective ways. No matter how tenderly they may regard one another, they have not yet made themselves to be absolutely unique and irreplaceable in their lives. A man and a woman do this, that is, make each other absolutely unique, when they "give" themselves unconditionally and irrevocably to one another in marriage. When a man and a woman give marital consent, that is, when they choose each other as husband and wife, they not only bring marriage into being but give to themselves a new identity. Through the act of marital consent[5] the man gives himself the identity of this particular woman's husband and the woman gives to herself the identity of this particular man's wife, and together they give themselves the identity of spouses. Through their own free and irrevocable choice they have established one another as absolutely unique in their lives, as irreplaceable and nonsubstitutable spouses.

Moreover, by giving themselves to one another in marriage husbands and wives not only acquire rights that nonmarried men and women do not have but also capacitate themselves to *do* things that nonmarried men and women are not capable of doing. Nonmarried men and women have the natural capacity, by virtue of their sexuality and their endowment with sexual organs, to engage in genital sex and through it to generate new human life. Yet they do not have the *right* either to engage in genital sex or to generate new human life. Although I cannot here show fully why they do not have the right to engage in genital sex,[6] I can briefly indicate why. The reason is that they have not, by their own free, self-determining choice, capacitated themselves to respect each other as irreplaceable and nonsubstitutable persons in their freely chosen genital acts. When unmarried men and women have sex, their genital act *does not unite two irreplaceable and nonsubstitutable persons* but rather *joins two individuals who are in principle replaceable and substitutable, disposable.* But human beings ought not to be regarded and treated as if they were replaceable, substitutable, disposable things.

Similarly, nonmarried men and women do not have the right to generate new human life precisely because they have not, through their own free choice, capacitated themselves to receive such life lovingly, nourish it humanely, and educate it in the love and service of God and neighbor.[7] Practically all civilized societies rightly regard as utterly irresponsible the generation of new human life through the random union of unattached men and women. Children have a right to the home that only a man and a woman united in marriage can provide. Deliberately to deprive them of this home is an injustice.

But husbands and wives have the right to an intimate sharing of life and love and to the marital act, whose nature will be examined more fully below. They have this right because they have capacitated themselves, through their irrevocable gift of themselves to one another in marriage, to respect one another as irreplaceable and nonsubstitutable spouses. When they choose to engage in the marital act, this act truly unites two irreplaceable and nonsubstitutable persons. Similarly, they have capacitated themselves to receive human life lovingly, nourish it humanely, and educate it in the love and service of God and neighbor, for by their free, self-determining choice to marry they have made themselves capable of receiving any new life that might be given to them and of providing it with the home to which it has a right and in which it can take root and grow under the loving tutelage of its own mother and father, persons who are not strangers to one another but uniquely *one flesh* in marriage.

An analogy may be helpful here. I do not have the right to diagnose sick persons and prescribe medicines for them. I do not have this right because I have not freely chosen to give myself the capacity to do these things. But doctors, who have freely chosen to submit themselves to the discipline of studying medicine and acquiring medical skills, do have this capacity and this right. Similarly, nonmarried men and women do not have the right to engage in genital sex or to generate new human life because they have refused to freely choose to give themselves the capacity needed to do these things well. But husbands and wives, by freely choosing to give themselves to one another irrevocably in marriage, have capacitated themselves to do what married people do, namely, to give themselves to one another in the beautiful act proper to marriage and to receive, in and through this act, the gift of new human life. The act proper to marriage is the "marital" act, which is an utterly unique kind of human act.

The marital act is not simply a genital act between a man and woman who happen to be married. Husbands and wives have the capacity to engage in *genital* acts, as do nonmarried men and women, because of their sexuality and endowment with genitalia. But they have the capacity (and the right) to engage in the *marital* act only because they are married, i.e., husbands and wives. The marital act, therefore, is more than a simple genital act between a man and a woman who just happen to be married. It is an act that inwardly participates in their marital union; it is an act inwardly participating in the "goods" of marriage, i.e., the good of steadfast fidelity and of exclusive marital love and the good of children. The marital act is, therefore, an act which is (1) open to the communication of spousal love and fidelity and (2) open to the reception of new human life. A genital act forced upon a wife by a drunken husband seeking only to gratify his sexual urges and unconcerned with her legitimate desires is a genital act, but it cannot be regarded as a true marital act.[8] Similarly, in my opinion, a genital act between husbands and wives that is deliberately made hostile to the reception of new human life, i.e., an act of contracepted intercourse, is also made to be *nonmarital* precisely because it is an act deliberately made inimical to one of the goods of marriage.[9]

The marital act, in other words, is by its own inner nature love-giving or unitive and life-giving or procreative. And it is such precisely because it is *marital*, i.e., an act participating in marriage and the goods perfective of it. The bond, therefore, that unites the two meanings of the marital act is the marriage itself. But "what God has joined together, let no man put asunder." It is for this reason, I believe, that there is an "unbreakable connection between the unitive meaning and the procreative meaning" of the conjugal act.

The marital act, as Pope Pius XII rightly said, is not "a mere organic function for the transmission of the germs of life." It is rather, as he noted, "a personal action, a simultaneous natural self-giving which, in the words of Holy Scripture, effects the union in 'one flesh' ... [and] implies a personal cooperation [of the spouses with God in giving new human life]" (Pius XII, 1951). Indeed, as Pope Paul VI put matters: "because of its intrinsic nature [*intimam rationem*] the conjugal act, which unites husband and wife with the closest of bonds, also makes them capable [*eos idoneos etiam facit*] of bringing forth new life" (Paul VI, 1968, n. 12).

In addition, one can rightly say that the marital act speaks the "language of the body." It beautifully embodies the personal, bodily integrity of the spouses. To see what this means, I think that some observations of John Finnis concerning personal integrity are relevant. According to Finnis,

personal integrity involves ...that one be reaching out with one's will, i.e., freely choosing, real goods, and that one's efforts to realize these goods involves, where appropriate, one's bodily activity, so that that activity is as much the constitutive subject of what one does as one's act of choice is. That one really be realizing goods in the world, that one be doing so by one's free and aware choice, that that choice be carried into effect by one's own bodily action, including, where appropriate, bodily acts of communication and cooperation with other real people – these are the fundamental aspects of personal integrity (Finnis, 1985, 45).

In the marital act a husband and wife are indeed freely choosing and realizing real goods in the world – their own marital union and new human life. Their own bodily activity is surely a constitutive subject of what they do; and cooperation with another is not only appropriate but necessary. The marital act is an utterly unique kind of human act; it is a collaborative, personal act carrying out the choice of the spouses to actualize their marriage and participate in the goods perfective of it.

B. Procreation vs. Reproduction

As we have seen, when human life is given through the act of marital union, it comes, even when ardently desired, as a "gift" crowning the act itself. The marital act is not an act of "making," of either love or babies. Love is not a product that one makes; it is a gift that one gives – the gift of self. Similarly, a baby is not a product inferior to its producers; it is, rather, a being equal in dignity to its parents. The marital act is surely something that husbands and wives "do"; it is not something that they "make." But what is the difference between "making" and "doing," and what bearing does this difference have on the issue of *in vitro* fertilization and embryo transfer, whether heterologous or homologous?

In "making" the action proceeds from an agent or agents to something in the external world, to a product. Autoworkers, for instance, produce cars; cooks produce meals; bakers make cakes, etc. Such action is transitive in nature because it passes from the acting subject(s) to an object fashioned by him or her (or them) and external to them. In making, which is governed by the rules of art, interest centers on the product made – and ordinarily products that do not measure up to standards are discarded; at any rate, they are little appreciated, and for this reason are frequently called "defective." Those who produce the products may be morally good autoworkers or bakers or cooks or they may be morally bad, but our interest in "making" is in the product, not the producers, and we would prefer to have good cakes made by morally bad bakers than indigestible ones baked by saints who are incompetent bakers.

In "doing" the action abides in the acting subject(s). The action is immanent and is governed by the requirements of prudence, not art. If the action is morally good, it perfects the agent; if bad, it degrades and dehumanizes her or him.[10] Moreover, we must keep in mind that every act of making is also a doing insofar as it is freely chosen, for the choice to make something is something that we "do," and this choice, as self-determining, abides in us. Thus, in choosing to bake a cake for someone's birthday, one is choosing to deepen the good of human friendship and is "doing" something good and making oneself to be, in this respect, a good person. Likewise, in choosing to make pornographic films, one is choosing to do something evil because it dishonors the dignity of human persons. There are, in other words, some things that we ought not to make, because choosing to make them is morally bad.

As we have seen, the marital act is not an act of making. It is rather an act freely chosen by spouses to embody their marital union, one open to the communication of a special kind of love, marital love, and to the transmission of new human life. As such, the marital act is an act inwardly perfective of them and of their life as spouses, the life of which they are co-subjects, just as they are co-subjects of the marital act itself. Even when they choose this act with the ardent hope that, through it, new human life will be given to them, the life begotten is not the product of their art but is a "gift supervening on and giving permanent embodiment to" the marital act itself (Catholic Bishops of England, 1983, n. 23). When human life comes to be through the marital act, we can say quite properly that the spouses are "begetting" or "procreating" new human life. They are not "making" anything. The life they receive is "begotten, not made."

But when new human life comes to be as a result of *in vitro* fertilization and embryo transfer – whether heterologous or homologous – it is the end product of a series of actions, transitive in nature, undertaken by different persons. The spouses "produce" the gametic materials which others then use in order to make the final product, the child. In such a procedure the child "comes into existence, not as a gift supervening on an act expressive of the marital union ... but rather in the manner of a product of a making (and, typically, as the end product of a process managed and carried out by persons other than his parents" (Catholic Bishops of England, 1983, n. 24). The new human life is "made," not "begotten."

But a child is not a product inferior to his or her producers and subject to quality controls (even if the choice is made not to apply these controls). It is, rather, as noted already, a person equal in dignity to his or her parents. A child, therefore, ought not to be treated as if he or she were a product. A child, therefore, ought not to be generated by *in vitro* fertilization, heterologous or homologous.

I believe that the reasons given thus far to show that it is not morally right to engender new human life outside the marital act can be set forth in the form of a syllogism, which I offer for consideration. It is the following:[11]

> Any act of generating human life that is nonmarital is irresponsible and violates the respect due to human life in its generation. But *in vitro* fertilization, whether heterologous or homologous, and other forms of laboratory generation of new human life are nonmarital. Therefore, these modes of generating human life are irresponsible and violate the respect due to human life in its generation (May, 1983).

In my opinion the major premise of this syllogism is supported by the first two lines of reasoning employed by the *Instruction* and by the considerations set forth previously regarding marital rights and capabilities, the marital act, and the generation of human life. Indeed, it is principally and precisely because human life is begotten nonmaritally when it comes to be through the acts of exploitative copulation by nonmarried persons or through acts of spousal lust and abuse that society regards, and rightly so, such modes of generating human life as irresponsible and reprehensible. In such acts, human life surely does not come to be as a "gift" crowning an act of spousal love. [Here I wish to make it clear that human life, no matter how generated, is itself a precious good; our concern is not with the goodness of the life generated, but with the ways chosen to generate the life.]

The minor premise surely seems verified in heterologous artificial insemination and *in vitro* fertilization. These procedures can hardly be considered as "marital" acts. But it is also, in my judgment, verified both in homologous artificial insemination and homologous *in vitro* fertilization. Although married persons have, *de facto*, taken part in these procedures, their *marital status* is not what capacitates them for collaborating in such procedures. Not only are these procedures ones that can in principle be carried out with the collaboration of nonmarried individuals, they are also procedures in which the *marital character* of spouses who *happen* to be involved in them is totally irrelevant as such. What makes the husband and wife capable of taking part in them is not their marital union and the act – the marital act – which both embodies this union and is made possible only by virtue of this union. To the contrary, what capacitates them to colloborate in these ventures is simply the fact that they, like *nonmarried* individuals, are the producers of gametic cells, ova and sperm, that other individuals then use to *produce* the new human life. Just as spouses do not generate human life *maritally*, i.e., precisely by virtue of their marriage, when this life is initiated through an act of spousal abuse, so too they do not generate human life *maritally* when they simply provide, through acts of making, other persons with gametic cells that can be united by those persons' acts of making.

IV. SOME MAJOR OBJECTIONS

In my opinion, the strongest and most straightforward argument against *in vitro* fertilization, both heterologous and homologous, and other forms of the laboratory generation of human life is the argument that these ways of generating new human life are forms of "producing" or "making" human life. Thus here I will first examine some objections that Catholic theologians have raised against this line of reasoning. In doing so, I will also set forth and respond to the critique that Richard A. McCormick, S.J., has raised against my claim that these modes of generating human life are nonmarital, a critique to which I referred in note 11 above.

McCormick is also a leading critic of the argument that the laboratory generation of new human life transforms the act of procreation into an act of re-*producing*. According to him, spouses who resort to homologous *in vitro* fertilization do not see this as the " 'manufacture' of a 'product.' Fertilization *happens* when sperm and egg are brought together in a petri dish." Citing William Daniel, S.J. (1986, p. 27), McCormick then continues: "The technician's 'intervention is a condition for its happening; it is not a cause' " (McCormick, 1989, p. 337).

Moreover, he notes, "the attitudes of the parents and the technicians can be every bit as reverential and respectful as they would be in the face of human life naturally conceived" (1989, p. 337). Indeed, in McCormick's view (and in that of some other writers as well[12]) homologous *in vitro* fertilization and embryo transfer can be considered an "extension" of marital intercourse, so that the child generated can still be regarded as the "fruit" of the spouses' love. While it is preferable, if possible, to generate the baby through the marital act, it is, in the cases we are concerned with, impossible to do this, and hence their marital act can be, as it were, "extended" to embrace *in vitro* fertilization.

Given the concrete situation, any disadvantages inherent in the generation of human lives apart from the marital act, so these authors reason, are clearly counterbalanced by the great good of new human lives and the fulfillment of the desire for children of couples who otherwise cannot have them. In this concrete situation, it is not unrealistic, so this criticism goes, to say that *in vitro* fertilization and embryo transfer is simply a way of "extending" the marital act.

In my opinion this criticism is based on rhetoric and not a realistic understanding of what is involved. Obviously, those who choose to produce a baby make that choice only as a means to an ulterior end. They may well intend that the baby be received into an authentic child-parent relationship, in which he or she will live in the communion of persons which befits those who share personal dignity. If realized, this intended end for the sake of which the choice is made to produce the baby will be good for the baby as well as for the parents. But, even so, and despite McCormick's claim to the contrary, the baby's initial status is the status of a product. In *in vitro* fertilization the technician does not simply *assist* the marital act (that would be licit), but, as Benedict Ashley, O.P., rightly notes, "*substitutes* for that act of personal relationship and communication one which is like a chemist making a compound or a gardener planting a seed. The technician has thus become the principal cause of generation, acting through the instrumental forms of sperm and ovum" (Ashley, 1990, p. 71). Moreover, the claim that *in vitro* fertilization is simply an "extension" of the marital act and not a substitution for it is simply contrary to fact. "What is extended," as Ashley also notes, "is not the act of intercourse, but the intention: from an intention to beget a child naturally to getting it by IVF, by artificial insemination, or by help of a surrogate mother" (Ashley, 1990, p. 72).

Since the child's initial status in *in vitro* fertilization is that of a product, its status is subpersonal. Thus, the choice to produce a baby is inevitably the choice to enter into a relationship with the baby, not as an equal, but as a product inferior to its producers. But this initial relationship of those who choose to produce babies with the babies they produce is inconsistent with and so impedes the communion of persons endowed with equal dignity which is appropriate to any interpersonal relationship. It is the choice of a bad means to a good end. Moreover, in producing babies, if the product is defective, a new person comes to be as *unwanted*. Thus, those who choose to produce babies not only choose life for some, but – does anyone doubt it? – at times quietly dispose at least some of those who are not developing normally.[13]

McCormick also claims that the meaning of the term "nonmarital" in the minor premise of the argument I developed at the end of Section III is "impenetrable." He writes: "In his [May's] own definition, it refers to an action of which a couple is 'capable' only by being spouses. But what is such an action? Surely not sexual union. For we could reword May as follows: 'What makes husband and wife capable of participating in such activities (sexual acts) is not their spousal union but the simple fact that they are beings who have sexual organs' " (McCormick, 1984, p. 102). It seems to me that McCormick simply fails to pay careful attention to my description of the "marital act" as distinct from a "sexual (or genital) act" between persons who merely *happen* to be husband and wife. I grant that husband and wife, like nonmarried individuals, are capable of "sexual acts" because they are endowed with genitalia, but my argument is that they are capable of engaging in the "marital act" precisely because they are spouses and that this act embodies their marriage and marital union. McCormick simply misses the point of my argument.

Thomas A. Shannon and Lisa Sowle Cahill (and other Catholic authors) take the *Instruction* to task for focusing on *individual acts* of the married couple and not on the marriage as a whole. According to them, "it is the committed love relationship of the couple in its totality that gives the moral texture both to their sexuality and to their subsequent roles as parents. It is from the wholeness of the relationship that their specific physical acts of sex and conception take their moral purpose" (Shannon and Cahill, 1988, p. 138). Thus, in their view, while ideally the child should come to be from the individual marital act of the spouses, what is crucial is that the child comes to be from their marital unity as a totality.

The problem with this argument and critique of the Vatican *Instruction* is that it simply fails to take seriously enough the *self-determining* character of our free choices and of human acts. We make ourselves *to be* the persons we are in and through the actions we freely choose to do every day. Our actions are not merely physical events that come and go, like the falling of the leaves. While there is a physical component in our actions, their moral core is the free, self-determining choice which they embody. And choices last, for they abide within the person, as a constitutive element of the person's moral character, disposing him or her to act in a similar way in similar situations unless another free, self-determining choice is made to repent of what one has freely chosen to do. In choosing to collaborate in the generation of new human life in the laboratory, spouses choose to bring new life into being as a product and not to receive it as a gift crowning their marital embrace. They are choosing to "make" a baby. They make themselves to be "baby-makers." But human babies ought not to be made.

V. CONCLUSION

And this brings me to the conclusion of this paper. *Human babies ought not to be made*. They ought not to be made because human babies are human persons. And human persons, I submit, are like "words" that God speaks. Human persons are the beings made in the image and likeness of God (cf. Gn. 1:28). As such they are his living icons, his "words." They are the *created words* that the *Uncreated Word* became and is precisely to show us how deeply God loves us. In the Nicene-Constantinople Creed we profess that God's Uncreated Word was "begotten, not made." Human beings, God's created words, ought, like the Uncreated Word who became and is one of them, to be "begotten, not made." They are begotten in the marital act; they are made in *in vitro* fertilization, homologous as well as heterologous. Begetting is a personal act that cannot, by its nature, be "delegated" to others. And, as Janet Smith perceptively observes, spouses can no more delegate to others the privilege they have of begetting human life than they can delegate to others the right they have to engage in the marital act (Smith, 1990, pp. 58–59). There are simply some things that no else can do for us, they are so personal.

Some, unfortunately, may think that the position taken by the Vatican *Instruction* and by me in this paper is heartless and cruel, insensitive to the agony experienced by married couples who ardently, and legitimately, desire to have children "of their own," and are not able to have them because of such factors as blocked fallopian tubes. The agony they suffer is real, and their desire is legitimate. Yet it is necessary to take very seriously the means proposed for alleviating these desires and for "helping" them to have "children of their own." If the means proposed are, as I have argued here, morally wrong because they offend the dignity of the child they ardently desire and also tear at the bonds, so crucial for human existence, between marriage, the marital act, and the generation of new human life, then they ought not to be employed. Moreover, *in vitro* fertilization and other modes of laboratory generation of human life are (1) very expensive and (2) not very successful. These procedures, moreover, do not address the underlying causes of the couple's infertility. They do not heal or treat a pathological condition. Rather, they treat human desires. But medicine, I submit, ought not to seek to treat human desires, which are infinite in number, but should seek to heal and cure human persons of pathologies from which they suffer. There *are* alternatives. Already, for instance, surgical reconstruction of the fallopian tubes has some success – greater than *in vitro* fertilization. Moreover, why would it not be possible to have tubal transplants, as we have kidney transplants? If medical science would expend as much energy and thought to curing the underlying pathological problems of infertile couples as it has on developing new "re*productive*" technologies, I am confident that many couples could be helped.

Finally, I do not claim that advocates of the laboratory reproduction of children are necessarily wicked, Faustian-like individuals seeking "to play God." But I do think that they are playing with dynamite and acting unwisely, for despite their noble goals they are in fact eroding the bonds, crucial for human existence, between marriage, the marital act, and the generation of human life, and they are also, as I have argued, reducing babies to the status of products. As Leon Kass, a perceptive observer of the human scene once observed, "folly is much harder to detect that wickedness" (Kass, 1972, p. 39). In my judgment the laboratory generation of human beings is foolish because it changes the engendering of human life from an act of marital self-giving love to one of technical reproduction. It "makes" rather than "begets" human life. And there is an enormous difference between these two ways of generating human life. Thus I believe that human civilization owes much to this Vatican *Instruction*. It simply seeks to remind us of the dignity that is ours as persons made in God's image and likeness, as his "created words," who, like his Uncreated Word, are meant to be begotten, not made.

NOTES

[1] It should be noted that many people, particularly in affluent Western democracies such as the United States, where contraception has become a way of life, are favorably disposed to the use of heterologous insemination and fertilization to help a childless couple have a baby, at least in some way, "of their own." Some, in fact, see the artificial generation of human life as "more human" than the "reproductive roulette" of generating children through sexual coition. A passage from the late Joseph Fletcher eloquently expresses this point of view:

Man is a maker and a selector and a designer, and the more rationally contrived and deliberate anything is, the more human it is. Any attempt to set up an antinomy between natural and biological reproduction . . .and artificial or designed reproduction . . .is absurd. The real difference is between accidental or random reproduction and rationally willed or chosen reproduction. . . .If it [the latter] is "unnatural" it can only be so in the sense that all medicine is . . .It seems to me that laboratory reproduction is radically human compared to conception by ordinary heterosexual intercourse. It is willed, chosen, purposed, and controlled, and surely these are among the traits that distinguish *homo sapiens* from others in the animal genus. . . .Genital reproduction is, therefore, less human than laboratory reproduction, more fun, to be sure, but with our separation of baby making from love making both become more human because they are matters of choice, not chance ('Ethical aspects of genetic controls: Designed genetic changes in man,' *New England Journal of Medicine*, 1971, 285, 781–782).

[2] Pregnancy reduction" is an expression used by some doctors who deliberately kill within the womb "excess" children who have been conceived *in vitro* and implanted in their mother's wombs to enhance the likelihood that at least one child will survive pregnancy. If, however, all the embryos implanted survive, serious problems would be raised. To cope with these problems the choice is made to kill off the excess number of children, for they are "unwanted."

[3] Joseph Boyle, Jr. has sought to formulate the principles underlying the reasoning of the *Instruction* in his essay, 'An introduction to the Vatican instruction on reproductive technologies', *Linacre Quarterly* (July 1988) 55, 20–28; reprinted as 'An overview of the Vatican's instruction on reproductive ethics', in *The Gift of Life: The Proceedings of a National Conference on the Vatican Instruction on Reproductive Ethics and Technology*, Marilyn Wallace, R.S.M. and Thomas W. Hilgers, M.D. (eds.) (Omaha, NE: Pope Paul VI Institute Press, 1990), pp. 19–26. References here are to Boyle's work as reprinted in *The Gift of Life*. Boyle says that the most important principles operative in the *Instruction* can be expressed in five propositions. "First, God makes human individuals in His own image and likeness, and He is directly involved in the coming-to-be of each new person. Second, the human person is one being, bodily as well as spiritual, so bodily life and sexuality may not be treated as mere means to more fundamental purposes. Third, every

living human individual, from the moment of conception, should be treated with the full respect due a person and so is inviolable. A human being is always a he or she, an I or you, never an object, a mere something. Fourth, sexual activity and procreation can be morally good only if they are part of marital intercourse. Fifth, in marital intercourse, love-making and life-giving should not be separated" (p. 20). Boyle observes that in general the *Instruction* does not attempt to establish these principles, although it does provide some argumentation to support the principle requiring that all human individuals be treated as persons from the moment of conception (p. 21).

Here I should note that I have previously addressed the subject of the laboratory generation of human life. See the following: ' 'Begotten, not made': Reflections on the laboratory generation of human life,' in *Pope John Paul II Lecture Series on Bioethics*, Francis Lescoe and David Q. Liptak (eds.), Vol. 1, *Perspectives in Bioethics* (Cromwell, CT: Pope John Paul II Bioethics Center, 1983), pp. 31–60; ' 'Begotten, not made': Further reflections on the laboratory generation of human life,' *International Review of Natural Family Planning* (1986), 10, 1–22; and 'Catholic teaching on the laboratory generation of human life,' in *The Gift of Life*, pp. 77–92. The present essay draws on some of the themes developed in these essays.

[4] John Paul II developed his "theology of the body" in a series of addresses from September 5, 1979 to November 28, 1984. These addresses have been published in four volumes in English. The first of these, called *Original Unity of Man and Woman: Catechesis on Genesis* (Boston: St. Paul Editions, 1981), offers readers a good introduction to this theology.

[5] On this see Vatican Council II, Pastoral Constitution *Gaudium et spes*, n. 48; see also John Lucas, 'The 'Vinculum conjugale': A moral bond,' *Theology* (1975) 78, 226–240.

[6] I have sought to establish the truth of this claim elsewhere. See, for example, my *Sex, Marriage, and Chastity: Reflections of a Catholic Layman, Spouse, and Parent* (Chicago: Franciscan Herald Press, 1981), chapter 5; "Sexual Ethics and Human Dignity," in *Persona, Verita e Morale: Atti del Congresso Internazionale di Teologia Morale (9–12 aprile, 1986)*, pp. 477–495, in particular pp. 488–489.

[7] Centuries ago St. Augustine rightly and wisely noted that one of the principal goods of marriage is children, who are to be received lovingly, nourished humanely, and educated religiously, i.e., in the love and service of God and neighbor. See his *De genesi ad literam*, 9.7 (PL 34.397).

[8] It is worth noting here what Pope Paul VI had to say in *Humanae vitae*, n. 13, where he explicitly stated that a "conjugal act" (using the expression simply to designate an act between a man and a woman who happen to be married) imposed by one of the spouses upon the other against the other's reasonable desires violates the requirements of the moral order.

[9] On this question, which cannot, of course, be taken up here, see Germain Grisez, John Finnis, Joseph Boyle, and William E. May, ' 'Every marital act ought to be open to new life': Toward a clearer understanding,' *Thomist* (1988) 52, 365–426; reprinted in the book by the same authors and John Ford, S.J., *The Teaching of "Humanae Vitae": A Defense* (San Francisco: Ignatius Press, 1988).

[10] Classic sources for the distinction between making and doing are: Aristotle, *Metaphysics*, Bk. 9, c. 8, 1050a23–1050b1; St. Thomas Aquinas, *In IX Metaphysicorum*, Lect. 8, n. 1865.

[11] I first proposed this syllogism in my essay, ' 'Begotten, not made': Reflections on the laboratory generation of human life,' in *Pope John Paul II Lecture Series in Bioethics*, ed. Francis J. Lescoe and David Q. Liptak, vol. 1, *Perspectives in Bioethics* (Cromwell, CT: Pope John Paul II Bioethics Center, 1983), pp. 31–30. Richard A. McCormick criticized this argument in his 'Notes on moral theology,' *Theological Studies* (1984) 45, 102, and his criticism will be taken up below, when I consider objections raised by Catholic authors to the reasoning found in the *Instruction*.

[12] On this see Thomas A. Shannon and Lisa Sowle Cahill, *Religion and Artificial Reproduction: An Inquiry into the Vatican "Instruction on Respect for Human Life"* (New York: Crossroad, 1988), p. 138.

[13] In the previous paragraphs I have paraphrased material developed by Germain Grisez, John Finnis, Joseph Boyle, and William E. May in ' 'Every marital act ought to be open to new life': Toward a clearer understanding.'

BIBLIOGRAPHY

Ashley, B., O.P.: 1990, 'The chill factor in moral theology', *Linacre Quarterly* **57.4**, 67–77.

Boyle, J., Jr.: 1990, 'An overview of the Vatican's instruction on reproductive ethics', in M. Wallace and T.W. Hilgers (eds.), *The Gift of Life: The Proceedings of a National Conference on the Vatican Instruction on Reproductive Ethics and Technology*, Pope Paul VI Institute Press, Omaha, pp. 19–26.

Catholic Bishops of England Committee on Bioethical Issues: 1983, *In Vitro Fertilization: Morality and Public Policy*, Catholic Information Services, London.

Congregation for the Doctrine of the Faith: 1987, *Instruction on Respect for Human Life in Its Origin and the Dignity of Procreation*.

Daniel, W.: 1986, '*In vitro* fertilization: Two problem areas', *Australiasian Catholic Record* **63**, 21–31.

Finnis, J.: 1985, 'Personal integrity, sexual morality, and responsible parenthood', *Anthropos* **1.1**, 43–56.

Fletcher, J.: 1971, 'Ethical aspects of genetic controls', *The New England Journal of Medicine* **285**, 776–783.

Grisez, G. et al.: 1988, ' "Every marital act ought to be open to new life": Toward a clearer understanding', *Thomist* **52**, 365–426.

John Paul II: 1980, 'General audience of 16 January 1980', in *Insegnamenti di Giovanni Paolo II* **3.1**, 148–152.

John Paul II: 1981, *The Original Unity of Man and Woman: Catechesis on Genesis*, St. Paul Editions, Boston.

John Paul II: 1982, 'Apostolic exhortation *Familiaris consortio*', *Acta apostolicae sedis* **74**, 81–191.

Kass, L.: 1972, 'Making babies: The new biology and the "old" morality', *The Public Interest* **26**, 28–56.

Lucas, J.: 1975, 'The "vinculum conjugale": A moral reality', *Theology* **78**, 226–236.

May, W.E.: 1981, *Sex, Marriage, and Chastity: Reflections of a Catholic Layman, Spouse, and Parent*, Franciscan Herald Press, Chicago.

May, W.E.: 1983, '"Begotten, not made": Reflections on the laboratory generation of human life', in F. Lescoe and D. Liptak (eds.), *Pope John Paul II Lecture Series*, Vol. 1, *Perspectives on Bioethics*, Pope John Paul II Center, Cromwell, CT, pp. 31–60.

May, W.E.: 1986, '"Begotten, not made": Further reflections on the laboratory generation of human life', *International Journal of Natural Family Planning* **10**, 1–26.

May, W.E.: 1987, 'Sexual ethics and human dignity', in *Persona, Veritae Morale (Atti del Congresso Internazionale di Teologia Morale (Roma: aprile 7–12, 1986)*, Citta Nuova Editrice, Roma, pp. 477–495.

May, W.E.: 1990, 'Catholic teaching on the laboratory generation of human life', in M. Wallace and T.R. Hilgers (eds.), *The Gift of Life: The Proceedings of a National Conference on the Vatican Instruction on Reproductive Ethics and Technology*, Pope Paul VI Institute Press, Omaha, pp. 77–92.

McCormick, R.A., S.J.: 1984, 'Notes on moral theology', *Theological Studies* **45**, 102.

McCormick, R.A.: 1989, *The Critical Calling: Reflections on Moral Dilemmas Since Vatican II*, Georgetown University Press, Washington, pp. 329–352.

Paul VI, Pope: 1968, Encyclical *Humanae Vitae*, Janet Smith (trans.), in Smith's *Humanae Vitae: A Generation Later*, The Catholic University of America Press, 1991, pp. 269–295.

Pius XII, Pope: 1951, 'Address to the Italian Catholic union of midwives; text in 1952', *Catholic Mind*, **50**, 61.

Pius XII: 1956, 'Discourse to those taking part in the second Naples world congress on fertility and human sterility, 19 May 1956', in *Acta apostolicae sedis* **48**, 470.

Shannon, T.A. and Cahill, L.S.: 1988, *Religion and Artificial Reproduction: An Inquiry into the Vatican 'Instruction on Respect for Human Life'*, Crossroad, New York.

Smith, J.: 1990, 'The vocation of Christian marriage as an approach to the bioethics of human reproduction', in M. Wallace and T.R. Hilgers (eds.), *The Gift of Life: The Proceedings of a National Conference on the Vatican Instruction on Reproductive Ethics and Technology*, Pope Paul VI Institute Press, Omaha, pp. 49–60.

Thielicke, H.: 1963, *The Ethics of Sex*, Harper, New York.

Verspieren, P., S.J.: 1987, 'Les fecondations artificielles. A propos de l'Instruction romaine sur le don de la vie', *Etudes* **366**, 615–632.

JEAN PORTER

HUMAN NEED AND NATURAL LAW

I. INTRODUCTION

The 1987 Vatican *Instruction on Respect for Human Life in Its Origin and on the Dignity of Procreation*, generally known as *Donum Vitae*, attempted to apply Catholic teachings on the moral parameters for sexual activity and procreation to new developments in reproductive technology (Congregation). In the course of doing so, it appeared to condemn any form of reproduction that separates reproductive activity from sexual intercourse, including the increasingly widespread practice of *in vitro* fertilization (IVF) making use of the gametes of husband and wife. This procedure is often referred to as the "simple case" for the moral evaluation of reproductive technologies, because it does not involve the use of the germ cells of third party donors, and therefore does not raise questions about the exclusiveness and integrity of the marital relationship. By condemning even this "simple" form of non-sexual reproduction, the Vatican was generally seen as reaffirming the claim that there is an unbreakable link between sexual activity and procreation, which was asserted clearly and forcefully in *Humanae Vitae*.

Donum Vitae has elicited an odd mix of positive and negative responses from Catholics and non-Catholics alike. Cardinal Bernardin begins a commentary on the Instruction by observing that "A *New York Times* editorial said 'thoughtful people' can welcome this 'considered set of views' (Bernadin, p. 3). In an editorial in the *Chicago Tribune*, Kenneth Vaux, associate professor and chief of ethics in medicine at the University of Illinois Medical Center, Chicago, wrote that the *Instruction* affirms a much-threatened normative value of the natural goodness and sacred mystery of birth. Regrettably, in its desire to preserve the deeply human nature of procreation, it plays down the salutary potential of science to ameliorate incapacity in the same procreative gift' (March 20, 1987)." In contrast to these generally positive reactions, however, we must consider the remarks of the noted moral theologian Richard

McCormick, who observes that "I have discussed this 'simple case' with physicians, moral theologians, healthcare personnel, married couples, and priests. Although my discussants are certainly not exhaustive, *no one* I spoke with accepts the Vatican's rejection of the 'simple case'" (McCormick, 1987, p. 8). At the end of his analysis of the Instruction, he adds his own dissent: "In summary, then, I find the congregation's analysis and reasoning on 'the simple case' unpersuasive. So do many others" (McCormick, 1987, p. 10).

Of course, the arguments of *Donum Vitae* cannot be a matter of indifference to Catholics, since this Instruction carries the authority of the Vatican, speaking through the Congregation for the Doctrine of the Faith. That does not mean that its judgements are so definitive as to preclude all dissent, but it does mean that within a Catholic context, questions of authority and institutional identity cannot be altogether absent from a consideration of this Instruction.

At the same time, even within a Catholic context, it is possible to bracket questions of authority in order to be able better to understand the arguments of this Instruction, taken on their own merits. Most of those who have either criticized or defended this document, even within the Catholic community, have done so, and that is also the approach taken in this essay. By bracketing questions of the authority of this document for Catholics, I do not mean to suggest that these questions are unimportant, but only that the merits of the arguments of the Instruction call for separate examination, independently, as far as possible, of the authoritative status of the conclusions of the Instruction.

What, then, are we to make of the conflicting responses to *Donum Vitae*? An examination of the instruction indicates that they can be traced, at least to a considerable extent, to tensions within the document itself. That is, the arguments behind the conclusions of this Instruction are unclear in certain key respects, which, on examination, reflect ambiguity or uncertainty concerning the moral principles that are at stake. This ambiguity, even uncertainty, is magnified in the responses to the document, in turn, because it is rooted in deeper tensions within the moral tradition shared by the authors of the document and many of its respondents. Hence, *Donum Vitae* is an important witness to the Catholic moral tradition, because it reflects tensions that are to be found throughout contemporary Catholic thought. For this very reason, it calls for serious study and careful response on the part of all those (not necessarily all Catholics) who have some stake in preserving and developing the wisdom contained in the Catholic moral tradition.

II. DONUM VITAE: ARGUMENTS AND CONCLUSIONS

In order to appreciate the arguments of *Donum Vitae*, it is necessary to note the *ad hoc* nature of this document. *Donum Vitae* was developed by the Congregation for the Doctrine of the Faith, in consultation with theologians, scientists, and physicians, in order to respond to questions put to the Congregation by Episcopal Conferences and individual bishops. As such, it does not attempt a comprehensive presentation, or much less, a theoretical development of the magisterium's teachings on sexuality and reproduction. It simply applies those teachings, with the authority of the Congregation, to questions that have been raised concerning new medical possibilities.

The introductory section summarizes the relevant principles as follows: "The fundamental values connected with the techniques of artificial human procreation are two: the life of the human being called into existence and the special nature of the transmission of human life in marriage" (Congregation, pp. 9–10). With respect to the first of these principles, the Congregation reiterates the familiar teaching of the magisterium that from the first moment of conception, the human entity called into existence by the fusion of the parents' germ cells is a human person in the full sense of the word, deserving of the same respect and immunity from undeserved harm as any other person. Based on this principle, the Instruction goes on to evaluate prenatal diagnosis, therapeutic procedures carried out on the embryo, and experimentation on human embryos and fetuses in the same way as the analogous procedures would be evaluated, if they were performed by physicians on children or adults. That is, in each case the procedure in question is said to be legitimate if and only if it is performed with the consent of the parents and for the benefit of the embryo or fetus itself. Procedures that are performed with a view to possibly aborting the fetus, or which would invariably damage or destroy the embryo or fetus, even a nonviable embryo or fetus, are condemned.

Similarly, the Instruction condemns any procedure that would destroy or damage embryos conceived *in vitro*, including the use of such embryos for experimentation, the deliberate destruction of "surplus" embryos, the freezing of embryos, and non-therapeutic genetic engineering designed to produce a child with desirable qualities. As the Instruction observes, this prohibition would be sufficient in itself to raise serious doubts about the procedure of *in vitro* fertilization, since as it is currently practiced, this procedure generally results in the destruction of some "surplus" embryos.

At the same time, it would be possible, at least in theory, to carry out the procedure of artificial insemination in such a way as to avoid the destruction of "surplus" embryos, or inordinate risks to the embryo destined for implantation. Furthermore, with respect to what is by any reckoning a less serious moral concern, it is also possible in theory to perform this procedure in such a way as to avoid requiring the husband to masturbate in order to procure a semen sample. Given these possibilities, and given the fact that homologous IVF by definition does not involve the use of either sperm or ova donated by third parties, is there anything *necessarily* morally problematic about this procedure?

In the judgement of the Congregation, homologous IVF, even when performed with all the safeguards noted above, is morally illicit for the same reason that the use of contraceptives is illicit. Quoting *Humanae Vitae*, the Instruction affirms the "inseparable connection, willed by God and unable to be broken by man on his own initiative, between the two meanings of the conjugal act: the unitive meaning and the procreative meaning" (this norm is generally referred to as the inseparability principle; Congregation, p. 26). It then goes on to draw the consequences of this connection for the question at hand: "Contraception deliberately deprives the conjugal act of its openness to procreation and in this way brings about a voluntary dissociation of the ends of marriage. Homologous artificial fertilization, in seeking a procreation which is not the fruit of a specific act of conjugal union, objectively effects an analogous separation between the goods and the meanings of marriage" (Congregation, p. 27).

At this point, the Instruction presupposes a certain understanding of natural law that has played an important role in Catholic moral analysis for centuries. According to this view, the natural, biological functions of the human organism have definite purposes, which can be recognized as such and which place moral parameters on human activity. Traditionally, the sexual function was seen within this framework as having only one primary purpose, namely, the procreation of children. However, in this century, Catholic theologians, and eventually, the magisterium itself, have acknowledged that the sexual act has two primary purposes, the unitive (that is, it serves to express and strengthen the personal relationship between the partners) and the procreative. In distinction from the earlier view, these purposes are seen as equally fundamental, but at the same time, they are also held to be inseparable. That is, any exercise of the sexual function that is deliberately closed to either of these goods is morally prohibited, because such an action would directly contravene one of the purposes built into the structure of the act of sexual intercourse. Such an action would not only be a violation of human nature, it would also be a transgression of the law of God, who is the author of human nature.

The reasoning of the Instruction goes beyond traditional natural law analysis in another way as well. Given the parameters of the traditional Catholic understanding of the natural law, it is easy to see how the inseparability principle could be applied to an act of sexual intercourse, in such a way as to lead to the conclusion that the use of contraceptives is problematic, because it violates one of the natural functions of the sex act. In offering this observation, I do not mean to suggest that this argument is necessarily conclusive, even granting the premises of this particular theory of the natural law, but at least it is straightforward enough. However, as a number of theologians have observed, even if we grant the inseparability principle, there is nothing logically compelling about the conclusion that procreation apart from the sexual act is morally illicit (McCormick, 1989, pp. 336–340). This principle rests on a particular analysis of the structure and the intrinsic purposes of the sexual act. But of course, in the procedure of artificial procreation, there is no act of sexual intercourse going on, to be tampered with or violated. Why, then, is it necessary to condemn this procedure together with, and under the same rubric as, the use of contraceptives?

The Instruction argues on the basis of an analogical extension of the inseparability principle, by which that principle is interpreted broadly as a general claim that the values of unitive love and procreation must be linked together, rather than narrowly, as a claim about the moral parameters set by the sex act considered in itself. In doing so, it goes beyond the arguments of *Humanae Vitae*, which does not address the issues raised by IVF at all, to return to the perspective of Pius XII, who similarly appealed to the natural link between sexuality, in its physical, emotional and spiritual functions as an expression of love, and procreation, to condemn any form of procreation apart from the sex act, as well as the use of contraceptives as a part of that act (McCormick, 1989, pp. 334–335). Thus, it goes beyond the logic of the traditional natural law analysis to offer a more existentially oriented reflection on the meanings inherent in the sexual love between spouses:

> Spouses mutually express their personal love in the "language of the body," which clearly involves both "sponsal meanings" and parental ones. The conjugal act by which the couple mutually express their self-gift at the same time expresses openness to the gift of life. It is an act that is inseparably corporal and spiritual. It is in their bodies and through their bodies that the spouses consummate their marriage and are able to become father and mother. In order to respect the language of their bodies and their natural generosity, the conjugal union must take place with respect for its openness to procreation; and the procreation of a person must be the fruit and the result of married love . . . Fertilization achieved outside the bodies of the couple remains by this very fact deprived of the meanings and the values which are expressed in the language of the body and in the union of human persons (Congregation, pp. 27–28).

The reference to the "meanings and values" of procreation, which are necessarily absent in IVF, prepares the way for the third argument that the instruction offers for condemning the procedure. That is, it is claimed that because IVF is aimed at bringing about conception through a technological procedure, it treats the child to be conceived as if she were a product. Thus, by resorting to this procedure, the parents and those who are aiding them in conception fail to respect the human dignity of the child to be conceived and born in this way:

> In his unique and irrepeatable origin, the child must be respected and recognized as equal in personal dignity to those who gave him life . . . He cannot be desired or conceived as the product of an intervention of medical or biological techniques; that would be equivalent to reducing him to an object of scientific technology. No one may subject the coming of a child into the world to conditions of technical efficiency which are to be evaluated according to standards of control and dominion (Congregation, p. 28).

What is the relationship of this argument to the appeal to the inseparability principle just discussed? While the Instruction is not altogether clear on this point, it does appear that IVF violates the dignity of the child to be conceived *because* it does not respect the "language of the body," and *therefore* regards the prospective child as a product of technology. This argument is offered immediately after the reflection on the language of the body and its meanings, and the two considerations are explicitly brought together:

> Only respect for the link between the meanings of the conjugal act and respect for the unity of the human being make possible procreation in conformity with the dignity of

the human person. In his unique and irrepeatable origin, the child must be respected and recognized as equal in personal dignity to those who gave him life . . .

In reality, the origin of a human person is the result of an act of giving. The one conceived must be the fruit of his parents' love. He cannot be desired or conceived as the product of an intervention of medical or biological techniques; that would be equivalent to reducing him to an object of scientific technology. No one may subject the coming of a child into the world to conditions of technical efficiency which are to be evaluated according to standards of control and dominion (Congregation, p. 28).

This passage suggests very strongly that what the Congregation finds objectionable about IVF is its intrusiveness at the very point of conception, which ought to be shrouded in the secrecy and mysteriousness of the processes of the body. This interpretation gains further weight from the fact that *Donum Vitae* does not condemn other medical procedures designed to facilitate conception and to safeguard the well-being of mother and child, provided that these are genuinely therapeutic in intent. Thus, it does not seem to be the technical nature of the procedure of IVF *per se* that is problematic, but rather, the fact that IVF is a technical procedure that results in conception apart from an act of sexual intercourse.

It would appear, therefore, that the argument that IVF violates the dignity of the child is intended to be taken as a further consequence of the inseparability principle, taken in the wider sense captured by the phrase, "the language of the body." Not everyone has read the Instruction in this way, however. At this point, it will be instructive to consider the interpretation of one of the most forceful defenders of *Donum Vitae*, William E. May, for whom the most basic argument against IVF is the argument that this procedure violates human dignity (May, this volume).

III. RESPONSES TO DONUM VITAE

In his " '*Donum Vitae*': Catholic Teaching Concerning Homologous IVF (the 'Simple Case')," (this volume, pp. 77–92) William E. May asserts that the argument that IVF fails to respect the dignity of the child to be conceived is the "most straightforward argument against resorting to the laboratory generation of human life," even while he acknowledges that "the other two lines of reasoning . . . illumine the wider issues concerning human existence raised by new reproductive technologies" (May, p. 77). In order to draw out these wider issues, he attempts to "probe the meaning of marriage and the relationship between marriage, the marital act, and the generation of human life" (May, p. 77).

In May's view, the distinction between marriage and other forms of sexual expression lies in the fact that married persons have mutually chosen to regard and treat each other as spouses, that is to say, as unique and irreplaceable partners in the acts of sexuality, procreation, and child rearing. Only within the context of marriage can persons engage in sexual activity in a way that is genuinely respectful of the unique individuality of the partner, because only in marriage is the spouse considered to be the only possible partner in sexuality and the upbringing of children. Nonmarried persons do not have a right to engage in genital sexual activity, because "they have not, by their own free, self-determining choice, capacitated themselves to respect each other as irreplaceable and nonsubstitutable persons in their freely chosen genital acts" (May, p. 78). For this reason, sexual activity outside the context of marriage involves regarding the partner as being "in principle replaceable and substitutable, disposable" (May, p. 79). In other words, sex outside marriage treats the partner as a means to sexual gratification or procreation, not as a unique end in herself or himself, thus violating the Kantian precept that rational humanity, whether in one's own person or that of another, should be treated as an end and never as a mere means.

We have been prepared for this thoroughly Kantian sexual ethic by May's assertion that the central argument against IVF to be found in *Donum Vitae* is the argument that the use of this technical procedure fails to respect the humanity of the child to be conceived. When he turns specifically to an analysis of that argument, the Kantian thrust of his analysis becomes still clearer. In his view, while it is of course legitimate to desire a child and to act on that desire, it is not legitimate, not consonant with the norm of respect for persons as ends in themselves, to attempt to bring a child into being through a technical procedure, which in effect places the child in the position of a product, rather than as a person deserving of respect as a unique human person. As he says, "When human life comes to be as a result of *in vitro* fertilization ... it is the end product of a series of actions, transitive in nature, undertaken by different persons. The spouses "produce" the gametic materials which others then use in order to make the final product, the child ... But a child is not a product inferior to its producers and subject to quality controls ... A child, therefore ought not to be generated by *in vitro* fertilization ... " (May, p. 83).

Both sexual acts and acts of procreation are thus to be evaluated by the norm of respect for persons, understood as equivalent to consistency with one's own freely given promises and to respect for the freedom and individuality of the other. This analysis of the morality of sexuality and reproduction thus gives pride of place to autonomy, both as determining the moral meaning of sexual relationships, and as calling for respect for the other as an actual or potential free moral agent.

Given this line of analysis, what becomes of the inseparability principle, as interpreted through the metaphor of the language of the body? In the first section, it was argued that within the Instruction itself, these are the primary arguments against IVF, and the appeal to the dignity of the child to be conceived is presented as a corollary of the inseparability principle. For May, on the other hand, the appeal to the dignity of the child to be conceived, taken in a thoroughly Kantian way, provides the key for interpreting the inseparability principle. After noting that of course, nonmarried men and women are naturally capable of procreation, he goes on to say that:

... nonmarried men and women do not have the right to generate new human life precisely because they have not, through their own free choice, capacitated themselves to receive such life lovingly, nourish it humanely, and educate it in the love and service of God and neighbor.[7]

But husbands and wives have the right to an intimate sharing of life and love and to the marital act, whose nature will be examined more fully below. They have this right because they have capacitated themselves, through their irrevocable gift of themselves to one another in marriage, to respect one another as irreplaceable and nonsubstitutable spouses. When they choose to engage in the marital act, this act truly unites two irreplaceable and nonsubstitutable persons. Similarly, they have capacitated themselves to receive human life lovingly, nourish it humanely, and educate it in the love and service of God and neighbor, for by their free, self-determining choice to marry they have made themselves capable of receiving any new life that might be given to them and of providing it with the home to which it has a right and in which it can take root and grow under the loving tutelage of its own mother and father, persons who are not strangers to one another but uniquely *one flesh* in marriage (p. 79).

... The marital act is, therefore, more than a simple genital act between a man and a woman who happen to be married . . . it is an act inwardly participating in the "goods" of marriage, i.e., the good of steadfast fidelity and of exclusive marital love and the good of children (p. 80).

... The marital act, in other words, is by its own inner nature love-giving or unitive and open to the transmission of human life or procreative. And it is so precisely because it is *marital* [emphasis in the original], i.e., an act participating in marriage and the goods perfective of it. *The bond, therefore, that unites the two meanings of the marital act is the marriage itself.* But "what God has joined together, let no man put asunder." It is for this reason, I believe, that there is an "unbreakable connection between the unitive meaning and the procreative meaning of the conjugal act (emphasis added, except where noted) (p. 80).

I have quoted May at some length because this passage clearly indicates the extent to which he grounds the moral meanings of human sexuality in free, self-determining choice. By the same token, it illustrates the tension between the Kantian emphasis on respect for persons, and other, older strands of the Catholic moral tradition, which can be found in *Donum Vitae* itself, as well as in May's defense of it.

For the authors of the Instruction, and for their predecessors in the Magisterium as well, the sex act is *naturally* both unitive and procreative in its effect, prior to any human choice about the meaning of the act. That is precisely why the sex act, as understood within official Catholic moral teaching, is appropriate only within the context of marriage, and that is also why there are constraints on the freedom of choice that partners have with respect to their sexual activities within marriage. For May, on the other hand, these goods are only present in sexual activity within the context of marriage, because it is precisely the partners' choice to marry that bestows those goods on marital sexual activity and links them together. In other words, for May human choice is finally determinative. When he confronts the question of why it is that human choice cannot also separate the goods of marriage, he falls back on what seems to be an appeal to a divine command, albeit an appeal that is taken out of its original Scriptural context: "what God has joined together, let no man put asunder."

May's defense of the arguments of *Donum Vitae*, in its Kantian emphasis on autonomous choice, reflects one important school of contemporary Catholic moral theology, which is also represented by Germain Grisez, John Finnis, Joseph Boyle and others (May, p. 81). As such, it illustrates both the strengths and the weaknesses of this school. On the one hand, May's insistence on keeping the norms of equality and respect for human dignity at the center of our moral evaluations is certainly consonant with the best of the Catholic moral tradition, and would meet with a sympathetic response on the part of many who stand outside that tradition.

On the other hand, May's appeal to autonomous self-determination as *the* source for the moral meanings of the sex act within marriage fits awkwardly, at best, with the Catholic moral tradition, in either its older or its more recent forms. It is hardly consonant with the older conception of the natural law, according to which the physical structure of at least some kinds of human actions is determinative of the moral value of those actions. And it fits awkwardly, too, with the more recent emphasis on the personal, affective, and relational dimensions of human sexuality, which are seen as inherent in sexual intercourse precisely because of the *natural* structure of this kind of act. In the words of Pius XII: "In its natural structure, the conjugal act is a personal action, a simultaneous and immediate cooperation on the part of husband and wife which by the very nature of the agents and the propriety of the act is the expression of the mutual gift which according to Holy Scripture brings about union 'in one flesh only' " (McCormick, 1989, p. 334).

At the same time, it should be noted that official Catholic teaching is in tension with itself just at this point. In its almost exclusive focus on the physical dimensions of sexuality, the earlier Catholic moral tradition undoubtedly left out a great deal that is central to any realistic moral evaluation of sexual behavior, and its final conclusions were distorted as a result. Nonetheless, the earlier teachings did at least have the value of clarity. It was very easy to know what the Catholic church was for and against in the area of sexual behavior, at least on the level of specific kinds of actions. The personal and relational dimensions of sexual behavior, on the other hand, resist this sort of clear-cut classification and evaluation. Once the magisterium introduced these considerations into its official teachings on sexual morality, it introduced an element that potentially works against the aim of providing clear-cut moral guidance.

A number of moral theologians, including, most notably, McCormick, have argued that if we take the new emphasis on the personal and relational with full seriousness, it would be difficult to sustain an absolute prohibition on IVF. These theologians readily grant that sexuality has unitive and procreative dimensions that should be held together, but what is critical, in their view, is that these dimensions be held together and expressed in the overall relationship between the partners, including its sexual expression over time. For these theologians, it is not necessary to insist that each individual act of sexual intercourse be open to procreation in order to preserve the quality of openness to new life that should characterize the relationship between spouses. Similarly, the act of seeking to conceive a child through IVF can express its parents' mutual, self-giving love, when that act is seen in the context of their overall relationship. As McCormick says, "If experience is our guide – and I think it clearly is not in the congregation's document – accepting medical interventions to overcome sterility between husband and wife is, or can be, precisely a concrete manifestation of their love. I have seen this repeatedly" (McCormick, 1989, p. 349). At the same time, those who argue in this way admit that the considerations raised by the Instruction are valid and should be given some moral weight. As McCormick says, "All artificial interventions, whether to promote or prevent conception, are a kind of 'second best' " (McCormick, 1989, p. 348). These considerations may lead us to raise serious questions about the use of IVF in a particular case, even to be reluctant to employ it. But for McCormick and those who follow his line of argument on this point, these negative considerations are not of such weight as to rule out the use of IVF altogether.

Admittedly, this way of approaching the moral issues raised by IVF cannot sustain the clear prohibitions of the Instruction, and for that reason, those who follow this line of argument are seen as "dissenting" theologians. But they can claim with justice that their dissent at the level of final conclusions is itself grounded in one strand of the official Catholic teaching on sexuality. Moreover, as we have just seen, the Kantian defense of the Instruction offered by May also involves a kind of "dissent" from official Catholic teaching on sexuality, in that it fails to take the natural and relational dimensions of human sexual behavior with the same seriousness as does the magisterium itself.

IV. ASSESSMENTS AND OBSERVATIONS

We should not be surprised to find that *Donum Vitae* generates such conflicting responses. After all, the Catholic moral tradition is a living tradition, and therefore, it is characterized by the internal tensions and conflicts that inevitably characterize a living tradition. Moreover, this tradition in particular is still in the process of assimilating the thoroughgoing intellectual and institutional changes inaugurated by the Second Vatican Council. Within this context, it is a sign of health that *Donum Vitae* should have generated very different responses, which can nonetheless all claim to represent some aspect of Catholic moral thought with fidelity. Correlatively, we should avoid the temptation to see any of these responses in terms of conformity to, or dissent from, the official Catholic teachings. They all represent some elements of that tradition and downplay or reject others. Indeed, the Instruction itself "dissents" from some aspects of an older Catholic understanding of the natural law and its application to sexuality, even as it affirms the more personalist orientation of *Humanae Vitae*.

Ultimately, one's position on the issues raised by IVF will depend on the position taken on other, more fundamental questions. How are we to understand the different dimensions of human sexuality, and indeed, of the human person, namely, the physical, the affective, the intellectual, and the volitional? How do these dimensions fit together and constrain one another in our choices and commitments? Once we have arrived at an anthropology that is theologically and philosophically satisfactory, what normative implications does that anthropology have? Or, to raise another set of questions, how are we to assess the significance of the moral law, for our personal lives, for our lives as members of church communities, for our standing before God? Is it of critical importance to identify and to conform to clear-cut rules, or should we take (some or all of) the norms of the moral law as ideals to be approximated, or as guidelines pointing us towards the kind of persons we should be and the kind of community life that we should try to foster? Of course, a series of questions does not an argument make. But it is apparent that it would take us well beyond the scope of this essay even to begin to address these questions.

For whatever it may be worth, my own sympathies lie with McCormick on this issue, although I would not agree with his analysis of IVF at every point. It seems to me that McCormick's personalist approach leaves room for incorporating the concerns and insights both of May and other Kantian moral theologians, and of the Instruction itself, whereas neither of the other two approaches can address these concerns so adequately. Furthermore, it does not seem to me that it is generally either possible or desirable to identify clear-cut, absolute prohibitions in the areas of sexual and reproductive morality, given the central importance of the affective and relational for this dimension of human life. However, I will not attempt to argue these views in any detail here, leaving them for another essay.

BIBLIOGRAPHY

Bernadin, Joseph: 1987, 'Science and the creation of life', *Responses to the Vatican Document on Reproductive Technologies*, The Catholic Health Association, St. Louis, pp. 3–6

Congregation for the Doctrine of the Faith: 1987, 'Respect for human life in its origins and on the dignity of procreation: Replies to certain questions of the day', United States Catholic Conference, Washington, D.C.

May, W.: 1996, 'Donum Vitae: Catholic teaching concerning homologous IVF', in K. Wildes, S.J. (ed.), *Infertility: A Crossroad of Faith, Medicine, and Technology*, Kluwer Academic Publishers, pp. 73–92.

McCormick, R.: 1987, 'Document is unpersuasive', in *Responses to the Vatican Document on Reproductive Technologies*, The Catholic Health Association, St. Louis, pp. 8–10.

McCormick, R.: 1989, 'Therapy or tampering: the ethics of reprductive technology and the development of doctrine', in *The Critical Calling: Reflections on Moral Dilemmas Since Vatican II*, Georgetown University Press, pp. 329–352.

JOHN W. CARLSON

INTERVENTIONS UPON GAMETES IN ASSISTING THE CONJUGAL ACT TOWARD FERTILIZATION

I. INTRODUCTION

In its Instruction *Donum Vitae* of 1987,[1] the Congregation for the Doctrine of the Faith offered "replies to certain questions" concerning interventions in human procreation, passed over other questions in silence, and, in concluding, called upon moralists to reflect deeply on the relevant teachings of the magisterium. In the years immediately following, a number of articles and scholarly volumes[2] responded, in one way or another, to the Instruction's call. However, important issues remain unresolved and/or require further clarification – even among those who accept *Donum Vitae*'s fundamental teachings. The present study treats one set of such issues, related to the moral acceptability of a range of interventions upon human gametes.

In part I, I summarize salient points in the Instruction and identify the issues to be treated. In part II, I discuss a series of technologies with sperm and ova, ordered in terms of their complexity and the extent to which processes of nature are supplemented. Finally, in part III, I consider relations between the preceding discussions and *Donum Vitae*'s stances on other interventions, in particular *in vitro* fertilization.

II. IDENTIFICATION OF THE ISSUES

At the beginning of the Instruction (1987, Introduction, 1), the Congregation recalls the church's "evangelical mission and apostolic duty" to present moral teachings "corresponding to the dignity of the person and to his or her integral vocation." These teachings relate to primary and fundamental rights of human persons. Their source, however, is not within the human simply as a species of nature. Following the long tradition of Catholic moral thought, as represented by St. Thomas Aquinas,[3] the Congregation affirms that "as she contemplates the mystery of the Incarnate Word, the church also comes to understand the 'mystery of man';" and, through her proclamation of the Gospel, the church reveals to persons their dignity and invites them "to discover fully the truth of [their] being" (1987, Introduction, 1).

The basic teachings of *Donum Vitae*, as they bear on the concerns of this study, might be summarized as follows.[4] In procreation, man and woman collaborate with God, who is directly involved in the coming-to-be of each human person. Every human individual should be treated with the full respect due a person from the moment of conception. Human persons are unified beings, bodily and spiritual; this affects the proper ends and thus the moral character of types of activity which, on a purely physical plane, we share with other animals. Among these are sexual activity and procreation, which are morally good as human ventures only insofar as they are part of marital intercourse. Finally, the lovemaking and lifegiving aspects of marital intercourse should not be separated.

In presenting and applying these general teachings, *Donum Vitae* refers to a number of earlier Vatican statements, including ones by Popes Pius XII, Paul VI and John Paul II. The inseparability mentioned above means, for example, that acts of generating new human life which *substitute* for or *replace* the conjugal act are morally unacceptable. However, quoting a 1949 discourse of Pius XII, the Congregation proposes (1987, II. B. 7) that "a medical intervention respects the dignity of persons when it seeks to *assist* the conjugal act in order to facilitate its performance or in order to enable it to achieve its objective once it has been normally performed."[5] The objective here intended is fertilization and the beginning of a new human life.

The notion of "assisting" the conjugal act was elaborated by Pope John Paul II in a greeting to fellow bishops who gathered to study issues in reproductive technology in light of *Donum Vitae*. After quoting from the Instruction, he writes (1988, p. xv) that it "is important to distinguish artificial fertilization ... from therapeutic techniques which aim at remedying the deficiencies of nature."

A variety of medical approaches to overcoming infertility would be acceptable to – and encouraged by – both the authors of the Instruction and John Paul II. He again writes (1988, p. xv) that "infertile couples ... have a right to whatever legitimate therapies may be available to remedy their infertility." Several such therapies are mentioned in *Catholic Identity in Health Care: Principles and Practice*, by Msgr. Orville N. Griese (1987, pp. 33–34); in addition to the manipulations of gametes to be discussed below, these include biochemical, psychological and surgical interventions.

Somewhat surprisingly, *Donum Vitae* itself proposes no examples of legitimate assistance to the conjugal act. Further, while it argues forcefully against non-therapeutic manipulation of human embryos, and against even homologous *in vitro* fertilization,[6] it proposes no general teaching regarding interventions upon human gametes.

One recent procedure, however, received approbation from all Catholic moralists and Church authorities who commented upon it, including Archhbishop Daniel E. Pilarczyk of Cincinnati. Called Low Tubal Ovum Transfer (LTOT), it was developed in Dayton, Ohio by Dr. David McLaughlin.[7] Seen as a likely remedy for blocked or diseased fallopian tubes, this procedure involved retrieving by laparoscopy a mature egg from the woman's ovary, and then reinserting it in the lower portion of the tube or in the uterus. The couple was encouraged to perform the act of marital intercourse both before and after the procedure, which was scheduled at the predicted time of ovulation.

As discussed in *Catholic Identity in Health Care*, the Pope John XXIII Medical-Moral Research and Education Center gave approval to LTOT. Other Catholic moralists, such as William E. May and Donald T. DeMarco, as well as Catholic physicians, such as the Australian Nicholas Tonti-Filippini, agreed.[8]

Somewhat similarly, but with the focus on the male reproductive system, a surgical procedure by which sperm are moved past a blockage in the epididymis, followed by a normal act of intercourse, appears acceptable to all Catholic commentators.[9]

Beyond such cases, however, there arise controversies about interventions with gametes. Certain authors who accept the principles of *Donum Vitae* approve more complex manipulations. Among the latter interventions are Tubal Ovum Transfer (TOT) or, more accurately, Tubal Ovum Transfer with Sperm (TOTS), developed by the same Dr. McLaughlin noted above; and a very similar procedure, Gamete Intrafallopian Transfer (GIFT), pioneered by Dr. Ricardo Asch, then of the University of Texas Medical Center in San Antonio. The procedures in question involve the retrieval of both ova and sperm, and the injection of them via catheter to a point as high as possible in the fallopian tube. Moralists at the Pope John Center studied the various protocols, formed an initial favorable judgment supposing certain concerns were met, then solicited and received support from an official in Rome.[10]

In light of such approvals, it is not surprising that a recent U.S. Congress Office of Technology Assessment report (1988, p. 365) takes it as a matter of settled Roman Catholic opinion that "gamete intrafallopian transfer" and other "interventions designed to augment the possibility of procreation through normal conjugal relations are morally licit." However, as Boyle points out (1988, p. 24), and as we shall see in detail in the inquiry below, the morality of GIFT and TOT in fact "is in dispute among moralists who accept [*Donum Vitae*'s] teaching."

In addition to the above procedures, and sometimes combined with them, are ones involving the treatment and storage of gametes, including their freezing or cryopreservation; and, either together with or separate from this, procedures by means of which gametes (sperm) from several conjugal acts are mixed and concentrated. About such interventions too there is controversy.

We should seek, therefore, an account of the range of morally acceptable interventions upon gametes which assist the conjugal act toward procreation. Further, to the extent that an array of such interventions appears justified, we should ask whether arguments in support of them would not also tend to justify certain interventions which *Donum Vitae* rejects; and, if so, we should see what other considerations ground the prohibition of the latter.

III. INTERVENTIONS UPON GAMETES: AN ORDERED INQUIRY

Let us begin by considering procedures which can be seen to fall within a narrow interpretation of "assisting the conjugal act ... to achieve its objective," then proceed to more complex and more difficult cases.

A. *Relatively Limited Interventions*

As already noted, two procedures involving the repositioning of gametes have received widespread approval from Catholic moralists: LTOT, in which an ovum or ova are moved past a blockage in the fallopian tube; and the somewhat comparable surgical procedure by which sperm are moved past a blockage in epididymis. In both of these cases, the medical procedures clearly are to be understood as means of "remedying the deficiencies of nature" and of rendering fruitful conjugal acts, which take place in normal fashion after the procedures are undertaken. (As indicated above, in the case of LTOT, the couple is encouraged also to engage in intercourse *before* the procedure is undertaken; this presumably is because sperm can survive in the female genital tract for a period of 48 to 56 hours, and thus sperm from the conjugal act prior to the transfer of the ovum might achieve the fertilization.)

Earlier in the century, prior to Pius XII's reflections, Catholic theologians had explored the meaning of and proposed cases of assisting the conjugal act by artificial means. One type of case which received widespread approval, according to Msgr. Griese (1987, p. 42), involved "retrieving the ejaculate of a particular act of ... intercourse and simultaneously projecting or propelling it closer to the cervical canal so as to enhance the possibility of fertilization". Even though this would disturb the normal motion of the sperm, a group of 21 theologians judged that if this were done in the context of a normal conjugal act, and with a view to assisting insemination, it would be a legitimate form of aid.

Griese notes (1987, p. 42n) that four of the theologians mentioned above made their approval contingent upon the supposition that the "seminal deposit was not withdrawn from the vagina." For these thinkers, it would seem, keeping the sperm from being *exteriorized*, as well as preserving their normal *route of travel*, were of crucial moral significance. Further, as suggested by the word "simultaneously" above, apparently all of the early 20th Century thinkers held that the intervention upon the gametes needed to be *immediate*, without temporal interval, in order to be morally acceptable.

It is interesting to apply these most restrictive moral interpretations to the modern cases discussed earlier, viz. LTOT and the movement of sperm past a blockage in the epididymis. It seems that the earlier thinkers' temporal concern would be fulfilled; for since the artificial assistance to the conjugal act occurs *prior* to the performance of the act, there would be no question of a temporal interval *between* the conjugal act and the act which assists insemination. Regarding the other two concerns, it seems, for the reason just mentioned, that the one related to gametes' route of travel *following* intercourse also would be fulfilled; and, arguably, that the one related to gametes' exteriorization would be preserved at least in substance – for both in LTOT and in the surgical movement of sperm the brief exteriorization of the gametes occurs *before* the conjugal act, and thus there is, in particular, no exteriorization of any sperm *from* a conjugal act.

By the 1950s, several Catholic medical moralists approved, supposing there were adequate justifying reasons, the temporary removal of the ejaculate from the vagina. For example, Father Otis Kelly, M.D., and the obstetrician Frederick L. Good, M.D., wrote (1951, p. 135) that semen can legitimately be removed in order to be "centrifuged to bring about a greater concentration of spermatazoa ... [This would be] not, strictly speaking, artificial, but ... rather an aid to natural insemination." More recently, as Griese notes (1987, p. 49n), the extra-vaginal treatment of sperm has been expanded to include other procedures, e.g., a process "involv[ing] the elimination of sperm-agglutinating and sperm-immobilizing antibodies by washing the sperm and then suspending them in a normal acellular seminal plasma". Following the procedure, the sperm are reinjected into the cervical canal, with the hope that they will better be able to survive the environment and reach the site of fertilization.

Can these latter cases be justified on Catholic principles? It could not strictly speaking be claimed that sperm here traverse an uninterrupted route on their way to meeting ova in the fallopian tube. Nor could it be claimed that there is no temporal delay. Further, of course, the techniques just mentioned all involve the exteriorization of gametes. Still, it is obvious that these interventions are done as an aid to the fruitfulness of the conjugal act. In addition, as Griese points out (1987, pp. 45–47), the *time interval need not be great* between the couple's loving act of intercourse and the reinjection of the sperm; and, as another commentator notes (Huber, 1988, p. 70), even in the normal case "the timing of marital intercourse and the timing of natural fertilization do not coincide." Moreover, the present procedures can be said to maintain, although admittedly with the addition of one or more steps, the *normal route and sequence* of the sperm and ova as they travel to the site of fertilization in the fallopian tube.

Thus, according to Griese and the moralists upon whom he draws for support,[11] the present interventions with gametes can be seen as appropriate instances of assisting the conjugal act toward its objective. Accepting this now widely held understanding, let us say that although there has been no official church pronouncement regarding these interventions, they indeed seem to fall within the range of what John Paul II has called legitimate "therapeutic techniques."

B. Issues Regarding TOTS and GIFT

Let us turn at this point to procedures currently generating much controversy among Catholic thinkers – Tubal Ovum Transfer with Sperm (TOTS) and Gamete Intrafallopian Transfer (GIFT). As described in Part I, the procedures are very similar; a principal difference in the standard protocols is that whereas in TOTS (which was developed at a Catholic hospital) sperm is retrieved from a conjugal act, in GIFT sperm is obtained by masturbation.[12] However, it is clear that the latter can be adapted to Catholic teaching on this point.

In light of the preceding analysis, a number of issues immediately arise. With TOTS and GIFT there not only is an interruption in the movement of sperm, as well as a temporal delay. The very route of the gametes' travel to the fallopian tube is altered, with the natural one being bypassed altogether. Further, the sequence or natural temporal order is also altered: sperm and ova now arrive at the site of fertilization abruptly and simultaneously. Finally, it is through manipulations by physicians and technicians that the gametes are made to arrive in this manner. Do these differences from the normal process of fertilization affect the ethical analysis of TOTS and GIFT?

Regarding the issue of a temporal delay, Griese points out that in this case as in preceding ones the interval between the conjugal act and the reinsertion of gametes may be relatively short. Indeed, he quotes medical studies to the effect that the transfer of gametes can take place "within an hour's time" of the preliminary conjugal act and the harvesting of ova; and he concludes that "this factor would provide increased validity of the claim of a *moral union* between that particular act of conjugal union and the repositioning of gametes" (Griese, 1987, pp. 47–48, emphasis added).[13]

Is the alteration in the manner of movement of sperm and ova to the fallopian tube of significance? According to Griese, it indeed would "enhance the symbolism" of assisting the natural teleology if current research into the possibility of delivering gametes by transhysteroscopy through the vagina and uterus should prove successful. For such would be "more in keeping with the physiological route and sequence of human fertilization". However, since even through laparoscopic repositioning the gametes "reach the *natural and teleological site* of human fertilization" (i.e., the fallopian tube), TOTS and GIFT as currently practiced can be morally acceptable (1987, pp. 43 and 49; emphasis added).

What about the fact that procreation achieved through the present procedures involves additional acts (beyond the conjugal act of the couple), performed by physicians and technicians?

One critic of TOTS and GIFT, Nicholas Tonti-Filippini, goes so far as to say (1990, p. 76) that "the direct causal connection between the conjugal act and the origin of a human life must be uninterrupted by any other human act"; that it is only in this way that "the right to a dignified origin ... can be preserved". Of course, this point, if accepted, would apply even to the earlier discussed procedures of washing and treating retrieved sperm and then reinserting them into the cervical canal. Slightly less severe and less general is the argument of another critic, Donald T. DeMarco. He argues (1988, p. 134) that in TOTS and GIFT "the technological involvement is clearly more prominent in the realization of conception than the sexual intercourse between the spouses". Still other Catholic moralists suggest that, as distinct from the earlier types of case, we here have medical personnel performing independent generative acts, or "acts which have ... moral unity and intelligibility apart from the marital act."[14]

Replying to such critics, Father Donald McCarthy of The Pope John Center appeals to what he takes to be a "consensus that we're not opposed to all 'additional acts'."[15] Further, he argues that here, as opposed to a genuinely "substitute fertilization," we are merely "trying to assure the *proper fulfillment of the natural role* of the sperm and ovum" – i.e., their meeting "in the Fallopian tube of the wife" (1988, p. 143, emphasis added).

At this point, let us consider Tonti-Filippini's remark about "the direct causal connection" between the conjugal act and fertilization. Some years ago, as recently reported, Father Bernard Lonergan, S.J. offered pertinent reflections in a different context. Discussing difficulties in the theoretical foundations of the traditional ban on artificial contraception, he pointed out that "the act of inseminating is not an act of procreating in the sense that of itself, per se, it leads to conception". The relationship is a statistical matter, and we should ask whether, "no matter what the circumstances, the motives, the needs, any deliberate modification of the statistical relationship must always be prohibited?" (1990, pp. 7–8). It would be an over-statement to say that, in the normal case, there is no causal relationship, or indeed no direct causal relationship, between insemination and conception. However, the two events are neither simultaneous nor unmediated by natural processes. Whether or not one agrees with Lonergan's own application of his point to contraception, it seems to apply very well to the present interventions upon gametes: their aim is to increase the statistical likelihood that conception will take place.

In the end, as I have argued in a prior study (Carlson, 1989, p. 534), it seems reasonable to say that procedures such as TOTS and GIFT *assist and supplement* the conjugal act, rather than replace it with a different generative act. The interventions of physicians and technicians can be understood as acts *subordinated* to the conjugal act, with *with the whole carefully sequenced toward the common goal of procreation*. On this view, the primary generative act would remain the loving act of the spouses; and there would be, in Griese's terminology, a "moral union" within the sequence of actions and events.

In light of the above – while recognizing that TOTS and GIFT remain controversial among Catholic thinkers – let us propose that these procedures indeed can be accommodated within the framework of traditional moral principles.

C. *Further Interventions Upon Gametes*

To this point, we have dealt with procedures involving sperm from a single conjugal act, and their repositioning, often with ova,[16] in the genital tract or directly in the fallopian tube. We now extend our inquiry to more complex and, to some thinkers, more controversial efforts in the treatment of infertility.

A first type of case involves the freezing of gametes. In the presence of certain medical indications, the long-term storage of ova is recommended. Dr. Marian Damewood notes (1990, p. 61) that "cryopreservation may be a solution to the serious problem which arises when the wife who desires to preserve fertility must have diseased ovaries removed or will undergo radiation or chemotherapy." Dr. David McLaughlin adds (1988, p. 61) that because hyperstimulation of the ovaries ordinarily precedes the removal of ova by laparoscopy, "researchers continue to pursue improved techniques to cryopreserve 'extra' eggs for use later in a natural cycle, thus avoiding exogenous hormonal manipulation." In a case similar to that noted by Damewood, Dr. Tonti-Filippini describes a situation in which sperm might be frozen: "A man who had become sterile from some form of therapy, such as for a carcinoma, might have had the foresight to freeze his own sperm (having been licitly obtained), prior to the therapy" (1990, p. 77).

Turning to moral concerns about cryopreservation, a first involves risk factors. Johannes Huber notes (1990, p. 70) that the technique of freezing oocytes "presents great difficulties. The question whether the spindle apparatus is destroyed by the freezing process is still a subject of controversy." Again, McLaughlin mentions the possibility of ova being "destroyed during the freezing/thawing process as intracellular ice crystals form" (1988, p. 61). All Catholic commentators would agree that in cases where the cryopreservation of gametes poses serious risks of damage to potential future embryos, the procedure cannot be morally employed.

Apart from this point, Father Donald McCarthy also alludes (1988, p. 144) to the freezing of sperm; somewhat surprisingly, he rejects it without explanation as one among a number of "morally objectionable factors." However, since among the other factors he mentions is the obtaining of sperm by masturbation, it may be that McCarthy's negative assessment of the freezing of sperm involves the assumption that the sperm is obtained in this way. As indicated in the above illustration from Tonti-Filippini, such is not necessarily the case.

Cryopreservation of male gametes of course involves a long-term separation between the performance of the conjugal act and the reinsertion of thawed sperm. Does this in itself make a moral difference? It might seem so to thinkers from The Pope John Center, in light their temporal account of preserving a "moral union," as seen in the comments of McCarthy's colleague, Msgr. Griese.

Further, it might appear that here there is a stronger claim than in standard TOTS or GIFT that we have a separate generative act by the physicians and technicians. Although, as we have seen, Tonti-Filippini describes a case in which he finds the freezing of sperm acceptable, it is significant that in his example the insemination may be accomplished by the husband himself as a supplement to an immediate conjugal act. In another place (1990, pp. 73–74), Tonti-Filippini supposes a different type of case – one in which, "five months later, the ova and sperm are thawed and transferred separately to the Fallopian tubes where a new life originates." Of this latter case, he remarks that "the technician performs a generative act which is a direct cause of the origin of a child if the procedure is successful." The idea suggested above that the technician's act can be subordinated to the conjugal act, with the act of the couple regarded as the primary generative act, is not discussed by Tonti-Filippini.

A number of other interventions with gametes are practiced and also should be considered in our account. Griese notes that in certain cases of oligospermia (inadequate concentration or motility of sperm), the husband's ejaculate can be collected, then "spun down" and concentrated. Additionally, such sperm can be fortified and supplemented with sperm collected from prior conjugal acts, or even with sperm gathered by other morally acceptable means – e.g., drawn directly from epididymis (see Griese, 1987, pp. 45–46).[17]

In assessing these various possibilities, let us first note *Donum Vitae*'s statement (1987, II. B. 4. a.) that procreation, to be morally good, must be "the fruit of a specific act of conjugal union." Can this phrase be applied to the present types of case? The authors from The Pope John Center, Msgr. Griese and Father McCarthy, are at pains to try to show that it can. Thus the former quotes approvingly another moralist who says that a "concentrated deposit of sperm," achieved as described above, "could be placed within the generative tract of the wife, immediately before a natural marital act, in order to mix with and fortify the husband's ejaculate" (Griese, 1987, p. 46). Thus, on this account, the immediate conjugal act would simply be "assisted" by sperm from acts which had gone before. Somewhat similarly, McCarthy says that "it's known only to the good Lord which sperm fertilized the egg": sperm from the present act, or sperm "that was added in a supplementary fashion" (McCarthy, 1988, p. 144).

It seems to me, however, that these accounts are somewhat strained; and that a more accurate description of the practice of sperm concentration would be to say that here *the results of several conjugal acts together combine* to enhance the prospect of fertilization. This account would appear to make such treatment for oligospermia unacceptable in light of the Instruction's remark about "a specific act." However, as I have earlier noted (Carlson, 1989, pp. 533–534), there is an important ambiguity in the Congregation's language about the conjugal act. For in addition to the phrase "*a* specific act" of conjugal union (1987, II. B. 4. a.), we find "the *act specific to*" husband and wife (1987, II. B. 4. c.), and "*the* specific act" of the spouses' union (1987, II. B. 4. a.). I take it that whereas the first phrase is restrictive as indicated above, the second would bear a broader interpretation and indeed could apply to an *accumulation* of sperm from a number of individual acts; and I take it that the third phrase is ambiguous between the first two senses.

If we accept the present account of sperm concentration as conceptually more satisfactory than those of The Pope John Center, it emerges that these final interventions upon gametes could be approved on Catholic moral principles only if one adopts an expansive understanding of language of the conjugal act.

IV. TECHNIQUES REJECTED BY *DONUM VITAE*

Although not our focus in this paper, a significant portion of the Instruction bears on heterologous methods of achieving fertilization (1987, II, A.). It is of little surprise to discover that the Congregation finds it impossible to regard a child conceived in this way as the "fruit and result of married love;" or that it rejects these practices as involving a complete "separation" of the procreative from the unitive.

However, if procedures as invasive and as technically complex as TOTS and GIFT can be justified, it may seem surprising, at least at first, that certain others of a *homologous* nature are rejected. One of these is artificial insemination (AIH). A principal reason for the rejection of this procedure is again related to the source of the sperm, which as the Congregation notes is usually by masturbation. Another important reason is that ordinarily artificial insemination substitutes for the conjugal act. Indeed, as a precise matter, *Donum Vitae* holds that AIH "cannot be admitted *except* for those cases in which the technical means is not a substitute for the conjugal act but serves to facilitate and to help so that the act attains its natural purpose" (1987, II. B. 6., emphasis added).

On the restrictive understanding of Donald DeMarco, there apparently in fact would be no case in which artificial insemination is acceptable: for AIH would constitute a "distinct act," one which is "extrinsic to marital intercourse and completely outside the two-in-one flesh unity which the conjugal act defines" (1988, p. 134). This objection, as we have seen, DeMarco takes to apply to TOTS or GIFT as well. But supposing, as we have above, that these procedures can be accepted, it would seem that AIH in relevantly similar circumstances, in a similarly supplementary role, also can be accepted. Further and more complicated questions arise about *in vitro* fertilization with embryo transfer (IVF-ET).

Although they hold great hope for some patients, TOTS and GIFT are only medically indicated, indeed only technically possible, on the supposition that certain aspects of the female reproductive anatomy are in place. If the fallopian tubes are absent or destroyed by disease, these procedures cannot be used to assist the infertile couple. It is, as Damewood notes (1990, p. 53), in this or a similar " 'end stage' of infertility therapy" that IVF-ET may be recommended. And, in spite of differences in medical indications and success rates,[18] some Catholic physicians and moralists (as well as many others) believe that these diverse procedures have ethically relevant features in common. Because of this it has been suggested that homologous IVF-ET might be acceptable on Catholic principles after all.

For example, it is reported that in early discussions at The Pope John Center, prior to the issuance of *Donum Vitae*, Father Edward Bayer suggested that "as long as we begin with a conjugal act, maybe we can even go so far as to have the fertilization in the laboratory" (see The Pope John Center, 1988, p. 176). Moreover, in a transcript of discussion included in the Center's *Reproductive Technologies, Marriage and the Church*, an unidentified Bishop makes the following observation: "I'm in dialogue with technicians who are doing IVF and who believe that they are in conformity with the Instruction. They believe still that IVF can be done in a moral way" (1988, p. 177). Further, both the present writer (1989) and Johannes Huber (1990) have recently urged that arguments in favor of GIFT as morally acceptable would tend to support IVF-ET as well.

Interestingly, at about the same time the more conservative Catholic writers DeMarco (1988) and Tonti-Filippini (1990) employed the same logic to the opposite effect – viz., that since IVF-ET involves a morally unacceptable separation of procreation from conjugal union, so also, by parity of reasoning, must TOTS and GIFT. Let us consider the relevant issues afresh.

As we have seen, TOTS and GIFT involve modification of spatial and temporal factors – what Griese calls the gametes' "route and sequence" – in the effort to assist fertilization. Thus if these interventions indeed are acceptable, it cannot be an argument against IVF-ET that it modifies such factors itself. Again, TOTS and GIFT involve acts in the service of procreation by physicians and technicians. If these can be regarded as properly subordinated to the couple's conjugal act, it would seem that, in relevantly similar circumstances (what *Donum Vitae* calls the "simple case"), the medical interventions in IVF-ET can be so regarded as well.

Moreover, it may be asked, if the route and sequence of the gametes can be adjusted, why not the *site* of their coming together? What moral significance attaches to the fallopian tube, as opposed to, say, a laboratory petri dish?

To the Instruction and its commentators, a principal answer is that if gametes are brought together in the laboratory, then, in Joseph Boyle's words (1988, p. 26), "the child in its coming-to-be is treated as an object, a thing and not a person." This point is elaborated by John Haas as follows (1990, p. 111): "The child is treated as ... a thing manufactured out of an egg and sperm subject to quality control and domination by others. Such a manufacture of a person is inappropriate to his innate and unassailable worth." And Donald McCarthy – who, as we have seen, accepts TOTS and GIFT – remarks that IVF "is a substitute fertilization ... with all the liabilities that come with subjecting human generation to a laboratory as opposed to human generation in the body of the mother" (1988, p. 143).[19]

The following reflection also might support the point. While both in TOTS and GIFT and in IVF-ET, researchers measure "success rates" in terms of resulting pregnancies, there is an important difference between the two. As an act in itself, an instance of TOTS or GIFT can properly be completed without any zygote ever being formed – just as in the natural case of insemination. But an instance of IVF-ET by definition involves the coming-to-be of a new human life; and thus it also involves the above-cited dangers of abuse and domination.[20]

But if the liabilities of laboratory generation can be minimized, so that the new life in its coming-to-be is subject to no greater risks than in natural fertilization, and if, further, each viable embryo is to be implanted in the uterus of the wife, *must* it be the case that innate human worth is not respected? The answer, it seems to me, is unclear.

Another tack taken by the Instruction involves reference to "the meanings and values ... expressed in the language of the body and in the union of human persons;" we read that a "fertilization achieved outside the bodies of the couple" is thereby deprived of these proper meanings and values (1987, II. B. 4. b). Expanding on this point, William May argues that the practice of IVF-ET "refuses to acknowledge the deep human significance of the personal gift, bodily and spiritual in nature, of husband and wife to one another that is aptly expressed in the conjugal act, a personal gift that is itself fittingly crowned by the gift of new human life" (1988, p. 112).

Although most of this paper's considerations have been in a natural or philosophical vein, May's description of new life as a "gift" recalls the religious context of our inquiry. This context is even more manifest in *Donum Vitae*'s reference, for example, to "the conjugal act wherein the spouses cooperate as servants and not as masters in the work of the Creator who is Love" (1987, II. B. 4. c.). The implications of this religious theme are set out by Msgr. Elio Sgreccia as follows: "From a theological point of view ... artificial procreation presents itself as severing the link of obedience between procreators [i.e., the couple] and creator; it implies the refusal of God's transcendent design" (1990, p. 128).

For the Catholic thinker, as these comments indicate, there is intertwined with and supportive of our natural moral insight a specifically religious source in Scripture and tradition. It perhaps should be said that, on its own, our natural moral insight leads to no definitive conclusion about homologous IVF-ET – especially if, as we have proposed, along with thinkers from The Pope John Center, TOTS and GIFT are to be accepted. It is interesting, on this score, to compare issues at end of natural life. Granted that one may morally withdraw extraordinary means of maintaining life, is it ever acceptable to engineer the time and manner of death? Or does such a practice necessarily involve a disrespect for the dignity of human life? Here, as at the beginning of life, it is difficult, I believe, to be confident of our natural moral insight, even regarding general precepts. And in both cases individual judgments are apt to be clouded by desires, hopes and fears, as well as by cultural prejudices. Thus, as St. Thomas Aquinas noted long ago (see *Summa Theologiae* I-II, Q. 94, arts. 4–5), the Catholic moralist is grateful for specific insight gained from the awareness of the supernatural origin and destiny of human life.

The implications, in a pluralist society, of this intertwining of natural and religious insight about procreation – especially the implications for public policy issues – are important to consider;[21] but they lie beyond the scope of the present paper.

NOTES

[1] As is usual with Vatican documents, the Instruction is commonly referred to by the first two words of its Latin text. Its full title in English is "Instruction on respect for human life in its origin and on the dignity of procreation, replies to certain questions of the day." The authorized English version was published in 1987 by the Pope John

Center; it later appeared, with commentary, in other volumes, e.g., Pellegrino et al. (eds.), *Gift of Life*, and Shannon and Cahill, *Religion and Artificial Reproduction*. For ease of reference, I shall, in quoting from the text, note chapter, section, and subsection.

[2] Among scholarly volumes, the following in particular might be noted: Pellegrino et al. (eds.), *Gift of Life*; The Pope John Center, *Reproductive Technologies, Marriage and the Church*; Shannon and Cahill, *Religion and Artificial Reproduction*; and an issue of *The Journal of Medicine and Philosophy* devoted to Infertility and the New Reproductive Technologies, Spicker (ed.).

[3] The classical locus is St. Thomas Aquinas' *Summa Theologiae* I-II, QQ. 90–94. For recent discussions of pertinent issues in this tradition, see the essays in Pellegrino et al. (eds.), *Catholic Perspectives on Medical Morals*. Also of significance is Pope John Paul II's *Evangelium Vitae* [*The Gospel of Life*] (Boston: Pauline Books and Media, 1995), which appeared after the present study was substantially completed. It might be noted that the 1995 encyclical does not add any new moral teaching regarding the issues to be discussed below.

[4] For an excellent account of the document, to which the present summary is indebted, see Boyle.

[5] The reference is to Pius' discourse to participants in the 4th Inernational Congress of Catholic Doctors, 29 September 1949, in *AAS* 41 (1949), p. 560. Emphasis added.

[6] "Homologous" interventions are those involving gametes, or products of gametic union, coming from a married couple. They are contrasted with "heterologous" interventions, which involve donor sperm, ova, or embryos. The present study will assume the former context throughout. *Donum Vitae*'s argument against homologous *in vitro* fertilization occurs in section II. B. 5 of the document. The greeting from Pope John Paul II, quoted above, repeats the rejection of this technique.

[7] For an account of the development of this procedure, at St. Elizabeth Medical Center in Dayton, and of Archbishop Pilarczyk's statement of approval, see Griese, pp. 43–44.

[8] See DeMarco, p. 126, who quotes May; and Tonti-Filippini, p. 70. Unfortunately, in spite of medical hopes and moral approvals, the LTOT procedure failed to result in any pregnancies; thus it was abandoned by its developer in favor of TOT, or TOTS, discussed below.

[9] See again N. Tonti-Filippini, p. 70.

[10] See the account by the Pope John Center's Rev. Donald G. McCarthy, p. 143, in which he quotes a letter from Msgr. Carlo Caffarra, head of the Pope John Paul II Institute for the Family in Rome.

[11] Griese cites Benedict Ashley, O.P. and Kevin O'Rourke, O.P., Gerald Kelly, S.J., Charles McFadden, O.S.A., Thomas O'Donnell, S.J., Arthur Vermeersch, S.J., and John C. Wakefield. See Griese, pp. 42–46.

[12] The favored method of retrieving sperm in the TOTS procedure involves the use of a perforated plastic condom-like pouch (which does not have the spermicidal qualities of a latex condom); its trade name is the Silastic Seminal Fluid Collection Device. See Griese, pp. 52–53. For further details about TOTS and GIFT, see the accounts of Drs. McLaughlin and Damewood. Although explicitly rejected by *Donum Vitae* (see II. B. 6), as well as traditional Catholic teaching, masturbation as a source of sperm *for reproduction by a married couple* is treated as possibly acceptable by at least one author who in other respects seeks to adapt reproductive technologies to the Instruction's principles. See Huber, p. 72.

[13] The whole matter of temporal relations in the events leading to conception needs to be carefully understood and expressed. At one point, the Congregation speaks of "the moment in which the spouses transmit life to a new person" (Congregation, II. B. 7). This seems a rather loose use of language, in light of the processes which actually occur in the normal case. See McCarthy p. 141.

[14] See the discussions by William E. May and by John Haas, in The Pope John Center, pp. 165–167. For further reflections by DeMarco, see his *Biotechnology and the Assault on Parenthood* (San Francisco: Ignatius Press, 1991).

[15] See The Pope John Center, p. 179.

[16] Griese also discusses a variation on GIFT involving sperm only, which he calls Sperm Intrafallopian Transfer or SIFT. See, pp. 49–50.

[17] The acceptability of drawing sperm directly from the husband's epididymis was proposed as early as 1921 by Father Arthur Vermeersh, S.J. See Griese, p. 46n.

[18] For a discussion of the relevant medical data, see again Damewood, pp. 53–60; and McLaughlin, pp. 58–59.

[19] *Donum Vitae* notes that as "regularly practiced, IVF and ET involves the destruction of human beings" (II. B. 5.) – i.e., the discarding of excess or defective embryos. That this is a permanent danger even as techniques advance is seen in the procedure, reported in the *New England Journal of Medicine* in September 1992, by which very early embryos can be tested for cystic fibrosis and other genetic diseases; those which test positively are not implanted. However, the "simple case" of IVF-ET being discussed here would involve no embryo wastage.

[20] It is worth noting that if there indeed is a morally relevant difference between TOTS or GIFT and IVF-ET, the situation is complicated by the development of an intermediate procedure called PROST (Pro-nuclear Stage Transfer). See McLaughlin, p. 61, and Tonti-Filippini, p. 69. Here the transfer occurs before syngamy, and thus arguably – even on the strictest view – before the completion of the coming-to-be of the new human life.

[21] For general discussions of such implications, see the essays in Pellegrino et al. (eds.) (1989), Part V; and in Pellegrino et al. (eds.) (1990), Part IV.

BIBLIOGRAPHY

Boyle, J.: 1988, 'An introduction to the Vatican instruction on reproductive technologies', *Linacre Quarterly* **55**(1), 20–28.

Carlson, J.W.: 1989, '*Donum Vitae* on homologous interventions: Is IVF-ET a less acceptable gift than GIFT?', *The Journal of Medicine and Philosophy* **14**(5), 523–540.

Congregation for the Doctrine of the Faith: 1987, *Instruction on Respect for Human Life in Its Origin and on the Dignity of Procreation, Replies to Certain Questions of the Day* (authorized English translation of *Donum Vitae*), The Pope John Center, Braintree, MA.

Damewood, M.D.: 1990, 'Current technology of *in vitro* fertilization and alternate forms of reproduction', in Pellegrino et al. (eds.), *Gift of Life*, Georgetown University Press, Washington, D.C., pp. 53–66.

DeMarco, D.T.: 1988, 'Catholic moral teaching and TOT/GIFT', *Reproductive Technologies, Marriage and the Church*, The Pope John Center, Braintree, MA, pp. 122–139.

Griese, O.N.: 1987, *Catholic Identity in Health Care: Principles and Practice*, The Pope John Center, Braintree, MA.

Haas, J.M.: 1990, 'The natural and the human in procreation', in Pellegrino et al. (eds.), *Gift of Life*, Georgetown University Press, Washington, D.C., pp. 99–114.

Huber, J.: 1990, 'Possible modifications of artificial fertilization techniques: biological considerations which may influence theological considerations', in Pellegrino et al. (eds.), *Gift of Life*, Georgetown University Press, Washington, D. C., pp. 67–72.

Kelly, O.M. and F.L. Good: 1951, *Marriage, Morals and Medical Ethics*, P. J. Kenedy and Sons, New York.

Longergan, B.F., S.J.: 1990, 'Letter to Father Ora McManus (1968)', in *Lonergan Studies Newsletter* **11**, 7–8.

McCarthy, D.G.: 1988, 'Response [to DeMarco]', in *Reproductive Technologies, Marriage and the Church*, The Pope John Center, Braintree, MA, pp. 140–145.

McLaughlin, D.S.: 1988, 'A scientific introduction to reproductive technologies', in *Reproductive Technologies, Marriage and the Church*, The Pope John Center, Braintree, MA, pp. 52–67.

May, W.E.: 1988, 'Catholic moral teaching on *in vitro* fertilization', in *Reproductive Technologies, Marriage and the Church*, The Pope John Center, Braintree, MA, pp. 107–121.

Pellegrino, E.D., Langan, J.P. and Harvey, J.C. (eds.): 1989, *Catholic Perspectives on Medical Morals*, Philosophy and Medicine Series, 34, Kluwer Academic Publishers, Dordrecht, Holland.

Pellegrino, E.D., Harvey, J.C. and Langan, J.P. (eds.): 1990, *Gift of Life, Catholic Scholars Respond to the Vatican Instruction*, Georgetown University Press, Washington, D.C.

Pope John Paul II: 1988, 'To my brother bishops from North and Central America and the Caribbean assembled in Dallas, Texas', in *Reproductive Technologies, Marriage and the Church*, The Pope John Center, Braintree, MA, pp. xiii–xv.

St. Thomas Aquinas: 1964, *Summa Theologica*, Blackfriars, McGraw-Hill, New York.

Sgreccia, E.: 1990, 'Moral theology and artificial procreation in light of *Donum Vitae*', in Pellegrino et al. (eds.), *Gift of Life*, Georgetown University Press, Washington, D.C., pp. 115–135.

Shannon, T.A. and Cahill, L.S.: 1988, *Religion and Artificial Reproduction*, Crossroad Publishing Co., New York.

Spicker, S. (ed.): 1989, *The Journal of Medicine and Philosophy*, issue devoted to Infertility and the New Reproductive Technologies, 14(5).

The Pope John Center: 1988, *Reproductive Technologies, Marriage and the Church*, The Pope John Center, Braintree, MA.

Tonti-Filippini, N.: 1990, ' "Donum Vitae" and Gamete Intra-Fallopian Tube Transfer', *Linacre Quarterly* **57**(3), 68–79.

U.S. Congress, Office of Technology Assessment: 1988, *Infertility: Medical and Social Choices*, OTA-BA-358, U.S. Government Printing Office, Washington, D.C.

CAROL A. TAUER

DONUM VITAE: DISSENTING OPINIONS ON THE "SIMPLE CASE" OF *IN VITRO* FERTILIZATION

I. INTRODUCTION

Within the international community of persons interested in biomedical ethics, a striking variety of issues related to *in vitro* fertilization are being debated. Even with regard to what is often called the "simple case," that is, fertilization of a woman's ova by her husband's sperm for transfer back to her uterus, the questions are myriad.

Does society impose an imperative of biological parenthood on married couples, so that they feel coerced to try whatever methods medicine may offer? Is this pressure a particular imposition on the woman, who is the one who must undergo the medical procedures, carry the risks, and expend the most personal energy in the process? Is she being particularly exploited if the couple's infertility is due to her husband's disfunction (for example, low sperm count or poor sperm motility) rather than hers?

Are all clinics which offer *in vitro* fertilization (IVF) qualified to do so? Are some of them fraudulent in their representations? Do couples understand that the success rate may be extremely low, varying from no successes in some clinics to about 20 percent in the very best ones? Should IVF be regulated and clinics licensed (Tauer, 1989, pp. 6–7)? Does social justice support offering expensive medical technologies like IVF in the United States, where about 40 million people lack coverage for basic health care? Or in a world where people can starve to death by the thousands in countries like Somalia?

Eighteen years of practice of IVF have laid to rest two fears often expressed in initial discussions: that the required manipulation of zygotes and early embryos would result in an increased proportion of birth defects, and that children born as a result of IVF would experience negative psychological consequences. But other effects, particularly the increase in multiple-gestation pregnancies, have raised questions unforeseen in early discussions (Tauer, 1989, p. 5). If fertility treatments (whether IVF or other) result in high risk multiple pregnancies, is one permitted to reduce the number of fetuses selectively on grounds that all would otherwise perish? Is it ethical to use measures that tend to cause twin or triplet pregnancies, since if carried to term, these births are disproportionately premature–with all the morbidity and financial costs associated with treatment of prematurity?

Another effect unforeseen earlier is the possibility of postmenopausal pregnancy, either with donated eggs or with a couple's own previously frozen embryos. Are there ethical limitations on the age at which one may choose to carry a pregnancy and become a mother? May one impose increased risks due to maternal age on a child-to-be?

In the United States, a two-decade moratorium on federal funding of research on *in vitro* fertilization has both scientific and ethical implications. Issues of safety, efficacy, and appropriate conditions for use of IVF have not been systematically investigated by the community of American scientists and physicians. Privately-sponsored research projects have relied on the good will and financial support of couples receiving infertility treatment. Without the regulation that accompanies government funding, ethical standards for research vary widely and are often questionable. For example, it has become common to pay persons who donate eggs and sperm to infertile couples, thus bringing about a commercialization of the entire enterprise.

In 1993 Congress passed the National Institutes of Health Revitalization Act, which essentially lifted the ban on federal funding of research involving *in vitro* fertilization. Before proceeding to fund proposals, NIH requested an outside panel, the Human Embryo Research Panel, to recommend ethical guidelines for research related to laboratory fertilization. Two decades of unanswered questions were put onto the table for discussion by the Panel and comment by the public.

With such a plethora of ethical questions related to IVF, it would be easy to argue that this procedure is a genie that should not have been allowed out of the bottle. A consequentialist could claim that, although the procedure may have good outcomes for individual married couples, it carries disproportionate social costs. The consequentialist could argue that, although IVF solves some problems for some infertile couples, fifteen years of experience demonstrate that it has raised so many unforeseen ethical problems that the disvalues outweigh the values.

In turning from these issues to an examination of the arguments on *in vitro* fertilization in *Donum Vitae* (1987), one is struck by the change of focus. Suddenly we are in a world where a small number of principles govern a narrow but clearly enunciated set of conclusions. The simple case of *in vitro* fertilization, involving only wife and husband, is declared illicit because it violates a traditional Catholic understanding of human nature, marriage, and sexuality. Formulations have an abstract quality about them. While statements constantly refer to the experience of persons and married couples, the validity of experience is measured against the abstract formulations rather than the reverse.

Many sections of *Donum Vitae* make significant contributions to the ongoing international bioethical debate. Cautions about surrogate motherhood and other third party involvements, about research and experimentation with early embryos (or preembryos), about human gene therapy for nontherapeutic purposes: all these concerns have been taken seriously by ethics commissions and governments worldwide.

A broadening of the discussion of the simple case of *in vitro* fertilization could have brought the Vatican into the pluralistic debate in a similar way. Questions posed in the initial paragraphs of this article show that IVF has raised a wide range of ethical issues for persons of all philosophic and religious viewpoints. Regrettably, the narrow focus chosen for the section of *Donum Vitae* on homologous IVF (the simple case) severely limits its influence.

My task in this paper, however, is to present a critique of what *Donum Vitae* does say about the simple case of IVF. I will pursue this task primarily by using the arguments and words of Catholic theologians who dissent from the document's reasoning and conclusions. Where pertinent, I will also incorporate comments from members of the broader community and will at times include my own observations.

II. EVENTS LEADING TO *DONUM VITAE*

The teaching of the Catholic magisterium on artificial fertilization goes back to 1897. In that year the Holy Office responded to the question, "May artificial insemination of a woman be done?" by answering *no*. It provided no reasoning, however, and theologians generally assumed that the condemnation was related to the use of masturbation for obtaining sperm (Curran, 1982, pp. 117–118). Even conservative Catholic moralists (for example, Arthur Vermeersch and Thomas J. O'Donnell) held that the morality of artificial insemination itself, apart from masturbation, was an open question until 1949 (Curran, 1982, p. 118; Curran, 1988, p. 79).

In 1949, however, in an address to the Fourth Congress of Catholic Doctors, Pope Pius XII condemned artificial insemination by husband (AIH) apart from the use of masturbation. Pius XII stated that AIH violated the divine plan for procreation because the natural conjugal act was not present (Curran, 1988, p. 79). He reiterated this position in 1951, and in 1956 he related the condemnation to the prohibition of contraception. In this address he also explicitly rejected "the experiments in artificial human fecundation *in vitro*" (cited in McCormick, 1989, pp. 334–335).

These pronouncements were considered by many to be early, tentative reflections on artificial fertilization. Highly-regarded Roman theologians like Marcellino Zalba, S.J., and Jan Visser, C.S.S.R., felt free to disagree with the Pope's conclusions, as did various theologians in other parts of the world (McCormick, 1989, pp. 335–336).

After the birth in England of Louise Brown, the first "test-tube" or IVF baby, the Ethics Advisory Board established in the U.S. issued its recommendations. The Board included a prominent Roman Catholic theologian, Richard McCormick, S.J., and heard testimony from other theologians, including Charles Curran. Discussion focused on possible disposal or wastage of embryos, consequences of IVF for family life, the potential for birth defects, and the permissibility of IVF research.

After conducting months of meetings and considering oral and written testimony from experts and from the public, the Board issued its recommendations. It approved IVF as ethically acceptable within the context of a marriage relationship, and supported government funding of research to verify the safety and improve the efficacy of the procedure (Ethics Advisory Board, 1979). McCormick's agreement with these recommendations was based on two considerations:

1. The wastage of human embryos that are discarded naturally in the gestational process appears to be over 60 percent, and so one may accept similar losses in trying to achieve pregnancy through IVF;
2. It would be irresponsible to offer IVF unless scientific research were simultaneously being conducted to determine whether it was causing any harm (Ethics Advisory Board, 1978–1979).

Once it was demonstrated that *in vitro* fertilization could actually produce a baby, hospitals and clinics worldwide began to include this procedure among treatments for infertility. A number of Catholic hospitals and universities, and other hospitals with largely Catholic staffs and clients, for example in Ireland, did the same. These institutions saw themselves as acting out of Christian charity to come to the aid of couples who were otherwise unable to have children. They interpreted their medical role as a support and extension of a loving marital union, not a replacement for it (Falise and Régnier, 1987, p. 63).

In expectation of a definitive statement on reproductive technologies, a number of bioethicists, theologians, and scientists offered consultative assistance to the Vatican. Richard McCormick published an open letter of recommendations in the Jesuit magazine *America* (1987a, pp. 24–28, 39). Michel Falise, rector of the Catholic University of Lille and president of the International Federation of Catholic Universities, requested discussion with Cardinal Ratzinger, but apparently received no response (Vacek, 1988, p. 113, n. 129).

The Sacred Congregation for the Doctrine of the Faith issued its *Instruction on Respect for Human Life in its Origin and on the Dignity of Procreation* or *Donum Vitae* on March 17, 1987. The actual publication caused a flurry of press comment and immediate response. Many bioethicists and theologians expressed their dismay at the reasoning and conclusions of the section on homologous IVF, or the simple case. Reaction was characterized well by two French bishops (of Nantes and of Rennes) who both referred to "a shock" (Marcus, 1987, p. 56; Jullien, 1987, p. 60). The archbishop of Rennes said that the document's rejection of homologous IVF would not be easy to understand, and at the time, he did not even attempt to make it understandable (Jullien, 1987, pp. 59–60).

After initial vigorous comment and discussion, however, interest in the Vatican's condemnation of IVF died down. Even among Catholic theologians, there is little recent material discussing the morality of the simple case. LeRoy Walters, Director of Kennedy Institute of Ethics, described it as "a stagnant issue" as long ago as 1984 (cited in McCormick, 1989, p. 331).

Because many Catholic institutions have halted their plans for *in vitro* fertilization programs as a result of *Donum Vitae*, while others have continued these programs and have asked for dialogue with the Vatican, it is incumbent upon theologians and bioethicists not to ignore the issue. In addition, we must be sensitive to the confusion experienced by Catholic married couples who must deal with the anguish of infertility at the same time that their Church prohibits them from using remedies which probably seem morally unquestionable to them.

In the following sections of this article, I will summarize the main points made by theologians in response to *Donum Vitae*, focusing on the section on homologous IVF.

III. GENERAL CRITICISMS OF METHODOLOGY

A. *The Consultation Process*

Many critics of the process used by the Congregation for the Doctrine of the Faith (CDF) described its consultation process as secretive, narrow, restricted to those who held a conservative position, purely clerical and male, non-ecumenical, and lacking in adequate scientific expertise (McCormick, 1987a, pp. 24–26; Vacek, 1987, pp. 112–114; Harvey, 1989, pp. 485–490). McCormick notes that bioethics in America is at least a decade ahead of Europe, so that directors of many European centers have come to North America to learn from our experience. Yet he knows of only one American, not a well-known bioethicist, who was consulted by CDF on the ethical issues discussed in *Donum Vitae*.

John Harvey's careful study of the consultation process (1989, pp. 485–490) confirms the speculations of other authors. He has used all available resources to identify those who may have been consulted, who contributed to discussions, and who actually wrote the document. But even he must still describe his report as "Speculations Regarding the History of *Donum Vitae*." Though Cardinal Ratzinger stated that the document was "the fruit of a vast consultation," and actually provided numerical figures (62 moral theologians and 22 other professionals), no list of these consultants' names is available. The reason given for this secrecy is to protect the consultants against pressure from lobbying groups. Harvey is able to name some persons who most likely had input at one stage or another of the process. He is also able to match the style and content of various sections of *Donum Vitae* to known works of several authors. He concludes his study thus:

> The possibility of a biased consultation process is serious indeed. . . .Consultation with U.S. experts in bioethics seems to have been only indirect. The consultants to whom CDF appears to have listened reflect only the conservative views favored by the highest

Vatican officials, not a consensus of Roman Catholic moral theologians worldwide. Father McCormick's concern that U.S. bioethicists were not consulted seems justified (1989, p. 490).

Internal evidence also indicates that research and consultation were narrow. Footnotes consist almost entirely of other pronouncements of the magisterium. Theologians and bioethicists are not cited, although they have produced an enormous amount of writing on these topics. Similarly, the studies and recommendations of professional and government commissions on bioethics appear to be ignored.

By early 1987 at least 85 major statements representing at least 25 different countries were available (Walters, 1987, p. 4). At that time LeRoy Walters developed a detailed analysis of fifteen of the most significant statements, mainly representing government commissions in the U.S., United Kingdom, Australia, France, Germany, Spain, and the Netherlands. These fifteen statements all agreed that homologous IVF was ethically acceptable (1987, pp. 5–7). The fact that *Donum Vitae* does not even acknowledge that consensus weakens its credibility.

While *Donum Vitae* is accurate in its references to scientific and medical procedures, it includes no documentation from those who are involved in research and practice with these procedures. Perhaps more importantly, it lacks any concrete empirical data on the experience of married couples, whether parents or infertile. While it often refers to such experience, references are either abstractions or assumptions that are not validated.

The lack of broad consultation is particularly disturbing in a document which claims to base its conclusions on natural law reasoning. Such reasoning depends on reflection on human experience, both as lived by ordinary people and as studied through the expertise of the scholarly disciplines. Lisa Cahill criticizes the document for its refusal to consider that "not all traditional articulations of the unity of sex, love and parenthood measure up to the experience of people whose lives are sexual, marital, parental and also Catholic Christian" (1987, p. 247). Richard McCormick, himself a natural law theologian, remarks that "continuing reflection on these matters departs from human experience at its peril" (1987c, p. 22).

B. The Assumption of Consensus

A second criticism, related to the consultation process, focuses on the role of consensus in the promulgation of *Donum Vitae*. The methodology of the instruction is "to appeal to a consensus on the meaning of marriage, sex, and parenthood" and to draw conclusions based on that presumed consensus (Shannon and Cahill, 1988, p. 114). Thus, the conclusions are expressed as if they also represent a consensus, at least among Catholic Christians, and there is also appeal to all people of good will. The problem is that the consensus which is assumed may not exist, and that there appears to be little effort to check it out.

While the Catholic Church is not a democracy, it has traditionally held that its doctrines and moral teachings represent the shared beliefs of its members. In *Lumen Gentium* the Second Vatican Council stated that "the body of the faithful as a whole anointed as they are by the Holy One cannot err in matters of belief" (Abbott, 1965, p. 28). Both Pope Pius IX (1854) and Pope Pius XII (1950) satisfied themselves through worldwide consultation that there was "universal agreement from the bishops down to the last member of the laity" before infallibly proclaiming two dogmas about Mary (cited in Harvey, 1989, p. 483).

In matters related to marriage, sexuality, and parenting it would seem even more necessary to have a consensus among Catholics, including the laity, before enunciating moral positions. It is in this area that lay persons rather than celibate and exclusively male clergy would seem to have particular expertise. But a review of responses to *Donum Vitae* shows not only that a consensus supporting the teaching on homologous IVF fails to exist. On this issue, there appears to be a clear consensus for the opposite position.

While preparing a review article on the *Instruction*, Edward Vacek surveyed a wide range of Christian ethicists and ethical committees. Most either had no moral problem with homologous IVF or approved it with qualifications (1988, p. 128). Richard McCormick reports discussing the simple case with physicians, moral theologians, healthcare personnel, married couples, and priests. He concludes, "Although my discussants are certainly not exhaustive, no one I spoke with accepts the Vatican's rejection of the 'simple case'" (1987b, p. 53).

Even theologians and bishops who accepted the teaching in deference to the magisterium recognized that it would be "a shock" and "not easy to understand" (Jullien, 1987, pp. 59–60). At least four Catholic universities (Nimegen, Lille, and the two Louvains) announced shortly after the appearance of the *Instruction* that they would continue providing *in vitro* fertilization (McCormick, 1987b, p. 53). The chief administrators at Lille publicly explained their moral stance:

We are open to accepting moral evidence of illicity. But in all conscience, we cannot say that we have that evidence today (Falise and Régnier, 1987, p. 63).

Representatives of two *in vitro* programs at hospitals in Ireland stated their intent to continue these programs, though one acknowledged that some staff members might withdraw their participation due to the Vatican ruling (Barron, 1987, p. 13).

Outside the Catholic community, the consensus in favor of homologous IVF is even more unified. As noted earlier, Leroy Walters' study of fifteen ethical commissions worldwide found that they unanimously agreed that IVF was ethically permissible. They were, however, critical of some other practices; for example, 70 percent rejected surrogate motherhood (1987, pp. 3–9). Baruch Brody studied the positions of America's religious communities on homologous IVF and concluded, "I know of no other religious community that shares [the Catholic Church's] view" (1990, p. 54).

If the natural law teaching of the *Instruction* is supposed to be consistent with the life experience of believing Catholics, and moreover appeal to all people of good will, then the level of disagreement with the Instruction's position on homologous IVF is significant.

C. Substituting Assertion for Argument

A third criticism of the methodology of *Donum Vitae* claims that the document tends to substitute assertion for reasoned argument. Often a principle is stated, followed immediately by a conclusion which applies that principle. Cardinal Joseph Bernardin, while supportive of the *Instruction* as a whole, notes that agreement on a principle may be quite different from agreement on its interpretation or application (1987, p. 23). He seems to be implying what others have said more directly: that the *Instruction* often moves from principle to conclusion without showing how it got there. Sometimes one could just as easily take the same principle and use it to come to an opposite conclusion.

Theologians complain that *Donum Vitae* "substitutes assertions for nuanced arguments," and that thinking Catholics will simply reject conclusions that are "devoid of arguments whose reasoning they can follow" (Cahill, 1987, p. 246; Shannon and Cahill, 1988, p. 115). Other theologians narrow the focus of this complaint. Richard McCormick, for example, says, "I find the congregation's analysis and reasoning on 'the simple case' unpersuasive. So do many others" (1987b, p. 55).

I was able to make a fresh reading of the *Instruction* after several years. I found that the general discussion of anthropology and of principles often led me to expect a conclusion different from the one which was then enunciated. I experienced not merely an absence of reasoned argument; what I experienced was a series of *non sequiturs*. It was like reading a student paper which seems to be leading you in one direction, and then suddenly ends up at a different, totally unexpected conclusion.

A typical example of this disjunction is related to the non-dualistic anthropology of *Donum Vitae*. The document must be praised for its attention to the human being as an integral body-soul composite, so that the relationship of marriage essentially involves both bodily and spiritual intimacy. We are constantly reminded of

> the unity of the human being, a unity involving body and spiritual soul. . . . The union [of the couple is] not only biological but also spiritual . . . (1987, II.B.4.b).

But the very next sentence asserts that "fertilization achieved outside the bodies of the couple remains by this very fact deprived of the meanings and the values" which a union of human persons should represent. The general principle as enunciated was leading me to go beyond the literal and physical, but in the application of the principle I was abruptly brought back to the physical aspect as central.

In Section IV, where I address theologians' criticisms of specific arguments in *Donum Vitae*, I will include other examples where a conclusion does not seem to follow from stated premises.

D. Discussion is Really About Contraception

Some theologians have proposed that, in the section on homologous IVF, the authors of *Donum Vitae* were really more concerned about contraception than about artificial fertilization. The document suggests that allowing either artificial insemination or *in vitro* fertilization would invalidate the interrelated moral analysis that prohibits contraception. From the point of view of tradition and authority, that possibility must be foreclosed. Charles Curran understands the view of those who find such revision unthinkable: "Could the Holy Spirit permit the teaching office to be wrong in a matter of such great import" (1988, p. 81)?

If the moral acceptability of the simple case of IVF logically entails the moral acceptability of contraception, then if one holds contraception to be unquestionably and unalterably illicit, one would also have to declare IVF illicit. No other arguments would be needed. For consider the following syllogism, using the abbreviations F and C (artificial fertilization and contraception):

If F is licit, then C is licit.
But C is not licit.
Therefore F is not licit.

Provided one accepts the two premises, the conclusion automatically follows.

One wonders whether this syllogism is not the basic argument underlying the reasoning in *Donum Vitae*. McCormick believes that "the instruction is much more about contraception than about the new reproductive technologies" (1987c, p. 25). If so, then it is an effort to maintain a prohibition which, in *Humanae Vitae*, was established more on the basis of tradition and authority than on persuasive reasoning. And then Patrick Verspieren, S.J., is correct in claiming that the "ultimate reasons" for CDF's condemnation of all artificial procreation "are to be found ... in the arguments of authority and in the positions taken by the previous popes" (1987, p. 615).

Arguments from tradition and authority, coupled with the principle of inseparability, support the second premise of the syllogism, that contraception is not licit. The principle of inseparability is generally taken to be the main support for the first premise also. In the following section, I will examine this crucial principle in detail.

IV. THE MORAL PRINCIPLE OF INSEPARABILITY

A. Standard Interpretations

Donum Vitae states that its moral conclusions are strictly dependent on the principles it enunciates. Regarding the simple case of IVF, there is really just one principle: the principle of inseparability (McCormick, 1987b, p. 54). Not only does inseparability form the basis for the rejection of homologous IVF, but it is also the principle which is thought to interlock the argument prohibiting artificial fertilization with the argument prohibiting contraception. It is the principle believed to support the first premise in the syllogism of the previous section: If F is licit, then C is licit.

The principle of inseparability is defined in *Donum Vitae*, quoting *Humanae Vitae*, as "the inseparable connection ... between the two meanings of the conjugal act: the unitive meaning and the procreative meaning" (1987, II.B. 4.a). This connection, willed by God and inscribed in the being of man and woman, expresses the nature of marriage and the necessary connection of the goods of marriage. These two aspects, unitive and procreative, may not morally be separated within the marriage relationship.

The magisterium has consistently interpreted this principle to apply to individual acts of intercourse as well as to the marriage as a whole. Inseparability was applied this way in *Humanae Vitae* in order to continue the Church's prohibition of contraception. Each act of intercourse must be open to procreation as well as expressive of a loving union. In *Donum Vitae*, a similar application is made to IVF:

Homologous artificial fertilization, in seeking a procreation which is not the fruit of a specific act of conjugal union, objectively effects an analogous separation between the goods and the meanings of marriage (1987, II.B.4.a).

Therefore IVF must similarly be prohibited.

Some theologians who have questioned this interpretation of the inseparability principle regarding IVF have also questioned the magisterial interpretation in relation to contraception. These authors maintain that the inseparability of the unitive and procreative aspects applies to the marriage as a whole. Each marriage must be a loving union which is open to procreation. But this may well be true even if the inseparability is not present in each individual act of intercourse. A couple who are open to procreation in general may still morally prevent conception at some times and during some specific acts of intercourse.

Analogously, a couple whose union in both loving and open to procreation, but who find themselves unable to procreate naturally, may use artificial fertilization. This procreation may be separate from the marital union expressed in any specific act of intercourse, but it is not separate from the loving union of their marriage as a whole (see McCormick, 1987, pp. 53–55; Curran, 1982, pp. 121–123).

These arguments suggest that the morality of artificial fertilization is grounded in the same arguments as the morality of contraception. Thus, these theologians would appear to accept premise one of the previous syllogism but deny premise two, and so deny the conclusion, that IVF is not licit.

B. An Alternate Approach

A more careful look at the magisterial definition and interpretation of the inseparability principle suggests a third line of argument, in which the magisterial prohibition of contraception would stand while homologous IVF could be permitted. Now the principle of inseparability refers to the unbreakable connection between marital union and procreation, between the unitive and procreative meanings and functions. But there is a *logical* ambiguity present here: the unitive/procreative meanings and functions of *what*? In the magisterial definition, does the inseparability of the unitive and procreative refer to the conjugal act, or does it refer to the marriage relationship? What is the *context* within which these two meanings are said to be inseparable?

This question may appear to be the same one discussed in the previous section, where the magisterium's application to each conjugal act was challenged by theologians who argued that it should more properly be applied to the marriage as a whole. But the disagreement described there is a substantive disagreement. It represents a difference of opinion on a philosophical and moral claim, an "ought." The question I am now raising is a purely logical issue.

In the language of the magisterium there appears to be a fallacy of ambiguity, specifically equivocation. At times official documents refer to separating "the two meanings *of the conjugal act*: the unitive meaning and the procreative meaning" (DV, 1987, II.B.4.a; emphasis added). But at other times the reference is to the marriage as a whole: IVF effects a "separation between the goods and meanings *of marriage*" (DV, 1987, II.B.4.a); emphasis added).

Depending on which statement is truly the magisterial understanding, two different moral conclusions follow. (It is logically incoherent to maintain both definitions of the inseparability principle simultaneously.)

In opposition to theologians who accept the second definition of inseparability, the magisterium has consistently held that the first definition is correct (while also invoking the second one on occasion). I will follow McCormick (1988, pp. 337–338) in referring to the first definition as the narrow sense of inseparability, the second as the broad sense. The magisterium, insisting on the correctness of the narrow sense of the concept, has applied it in its prohibition of contraception. In *Donum Vitae*, it has claimed that the concept analogously applied also prohibits homologous IVF.

But this is logically impossible. For the narrow sense of inseparability is about the unitive and procreative meanings of the conjugal act. And as William Daniel, S.J., has aptly noted:

> All we learn from this passage [of *Humanae Vitae*], however, is that *if the conjugal act is performed* it should have these qualities....This does not necessarily imply that if there is to be procreation it should be by means of a unitive act (1986, p. 31).

We can go beyond Daniel's claim that the arguments of *Humanae Vitae* do not necessarily imply the prohibition of IVF. The narrow sense of the concept of inseparability does not in any way prohibit IVF.

Consider the following syllogism, where the first premise states the inseparability principle in its narrow sense, S refers to the act of sexual intercourse, and C refers to the connection of the unitive and procreative aspects:

If S occurs, then C is morally required.
But it is not the case that S occurs.
[Then any conclusion regarding C will validly hold.]

This syllogism has a first premise whose antecedent is false if there is no act of sexual intercourse. Hence no conclusion can be drawn as to the moral requirement regarding C.

The narrow sense of the principle of inseparability thus entails no conclusion about the moral acceptability of IVF. The authors of *Donum Vitae* must have realized this fact, since they slide over to the broad sense of inseparability in condemning IVF. Here they invoke the "separation between the goods and meanings of marriage" (DV, 1987, II.B.4.a.). But this slide is a simple fallacy of equivocation. One cannot have it both ways. As G.B. Guzzetti notes, the inseparability of the unitive and procreative within the conjugal act may not be transferred to an inseparability between the conjugal act itself and the generative process (cited in McCormick, 1988, p. 336). And this is precisely what the authors of *Donum Vitae* have done.

The magisterium may wish to adopt the broad sense of the principle of inseparability. Then they would be in accord with the opinions of most dissenting Catholic theologians, as well as the other religious traditions which use the concept of inseparability to express their theology of marriage. The broad sense would presumably imply moral acceptance of contraception in at least some circumstances. It could no longer be considered to be intrinsically evil at all times.

It is unclear whether the broad sense of inseparability could then be used to prohibit homologous IVF. Since most theologians who question the reasoning of *Humanae Vitae* and *Donum Vitae* argue against magisterial reliance on the narrow sense of inseparability, the *Instruction's* occasional invocation of the broad sense of the concept merits further examination.

V. OTHER ARGUMENTS

While the centerpiece of *Donum Vitae's* argument against IVF is the principle of inseparability, it does utilize other arguments to support its position. Many of these arguments have received comment from theologians and bioethicists. I will quote several passages from the document and then summarize responses to each one.

> The one conceived must be the fruit of his parents' love. He cannot be desired or conceived as the product of an intervention of medical or biological techniques (1987, II.B.4.c).

These two sentences are juxtaposed in *Donum Vitae* as if the second followed from the first. But there is no reason why the use of medical techniques to achieve fertilization should contradict the love of the parents. There are thousands of acts of love in marriage, and the sex act is only one of them. There is no reason to think that the use of IVF would be an unloving act. It may be the supreme example of the love of husband and wife, as they sacrifice together and work cooperatively in this process with its frustrations, hardships, and demands. The joint pursuit of treatment for infertility can be an example of enormous self-giving (Vacek, 1988, pp. 115–116; McCormick, 1987c, pp. 22, 25).

Some authors have noted that the availability of alternate reproductive techniques may actually free the act of intercourse "from frustrating preoccupation with vain attempts at procreation and allow it once again to unite the marriage partners in love" (W.B. and P.W. Neaves, pp. 13–14, 17). In this regard it is essential to consult the experience of married couples who are facing infertility and undergoing treatment for it. One couple wrote thus of their experience at the Norfolk IVF clinic:

> We were immensely impressed with the great concern ...and respect for human life exemplified by the doctors and staff.... They are there to assist you from a medical

standpoint; the couple provides the atmosphere of bonding and love necessary for the creation of a child (cited in McCormick, 1987c, p. 22).

Another passage from *Donum Vitae* focuses precisely on the professional persons mentioned by this couple:

> Homologous IVF ...is brought about ...through actions of third parties....Such fertilization entrusts the life and identity of the embryo into the power of doctors and biologists and establishes the domination of technology over the origin and destiny of the human person. Such a relationship ...is in itself contrary to ...dignity and equality (1987, II.B.5).

Many commentators have observed that the word choice in this passage is negative and harsh. Why use phrases like "the power of doctors" and "the domination of technology" when one could just as well refer to doctors and technology as helping and serving human ends and human needs? It is only because biased language was chosen that "power" and "domination" are in opposition to "dignity and equality." If the professional-client relationship were described in a different way, there would be no reason for this relationship to constitute a denial of the dignity and equality of the persons involved.

All medical treatment involves some lack of symmetry in the relationship. One of the disvalues of going to the doctor or being in a hospital is that one loses some independence and may be subjected to procedures that are discomfiting and embarrassing. But we do not usually describe them as "contrary to" human dignity.

Perhaps CDF sees the beginning of human life as unique, therefore uniquely free of the intervention of medical technology. But what about the end of human life, also a unique event? The intervention of medical technology can be extremely intrusive at the time of the ending of life. While the Catholic Church permits refusal of such technological intervention in the dying process, it does not prohibit its use.

And even with respect to the beginning of life, there are other situations where the intervention of medical technology is much more intrusive toward the child than IVF is. Consider the situation of a high risk delivery, an emergency cesarean section, a premature newborn whisked off to a different hospital which has a neonatal intensive care unit, and weeks if not months of the most sophisticated medical and technological procedures and life supports. This scenario would seem much more than IVF to represent domination by the medical system over the origin and destiny of the child. For one thing, the loving parental involvement is proportionately much less than with IVF.

Human experience of homologous IVF, coupled with comparison to other uses of medical technology which enhance human life, belies the claim of *Donum Vitae* in the passage quoted.

In a third passage, the *Instruction* states that in IVF, "The generation of the human person is objectively deprived of its proper perfection" (1987, II.B.5). The proper perfection of human reproduction results from a loving act of intercourse, engaged in by married persons, which enables them to join as co-creators with God of a new human being. This description represents a sort of ideal for the continuation of the human race. But as several commentators have noted, why is an act "deprived of its proper perfection" thereby morally illicit? All medical treatment provides alternatives when a function is deprived of its proper perfection.

McCormick characterizes this concern of the *Instruction* as an aesthetic or ecological concern (1987b, pp. 54–55). He acknowledges that artificial interventions into the reproductive process are "second best," and involve certain disvalues. But he is perplexed as to why this aesthetic-ecological issue has been elevated into "an absolute moral imperative."

Besides commenting on specific arguments of *Donum Vitae*, authors have criticized the weight given to various ethical concerns. Practices such as surrogate motherhood and nontherapeutic genetic manipulation seem to present ethical problems far more serious than homologous IVF. It is true that the *Instruction* adverts to this fact, stating that the simple case of IVF does not have "all that ethical negativity found in extra-conjugal procreation" (1987, II.B.5). But it still is strictly prohibited. This tendency to prohibit everything led Charles Krauthammer to comment, "When everything is prohibited, nothing is," and Daniel Maguire to lament, "The Vatican is squandering its moral authority" (cited in Goldman, 1987, p. 13).

Even with regard to homologous IVF itself, the document's narrow focus on the inseparability principle and on technological intervention leads to a neglect of other debatable aspects of this practice. Bishop Lehmann of Mainz believes that the treatment of "extra" embryos which may result from *in vitro* fertilization "has not been clarified at all." He also calls attention to "the unconditional desire to have a child" (1987, p. 62), which may drive people to desperate measures. Is the Church perhaps partly responsible for conditioning women to believe that biological motherhood is essential to their natures as women? In the early days of *in vitro* programs, the BBC interviewed a group of women being treated for infertility. One woman told how everything she had tried in her life had failed: school, work, personal and family relationships. But she always consoled herself with the thought, "At least I can be a mother." When she found that she could not physiologically do that, her life became completely meaningless and she was driven to overcome her infertility. This sort of desperation is a moral concern.

Other authors raise issues of social justice. In some countries *in vitro* fertilization is available only to those who are fairly affluent; in others it is subsidized by national health plans paid for through taxation. In either case it raises questions of justice in the allocation of the scarce health care resources that are available (Cahill, 1987, p. 247; Canadian Conference, 1991, p. 635).

The equality of women and men is an issue which is also overlooked in *Donum Vitae*. With some reproductive technologies, the burden falls particularly heavily on women. It has often been pointed out that surrogate motherhood has the potential to exploit women, particularly poor and disadvantaged women who may have an economic incentive to agree to act as gestational vessels (Cahill, 1987, p. 247). But other technologies also have the potential for exploitation of women, who may feel pressured by society or by their partners to undergo whatever rigors are necessary to become pregnant. Feminists have been particularly vocal in pointing this out (see Stanworth, 1987; Rothman, 1989; and the serial *Issues in Reproductive and Genetic Engineering: Journal of International Feminist Analysis*). Recent studies show that male infertility is a significant cause of failure to conceive, and that IVF is one of the most effective treatments to overcome many forms of male infertility: low sperm count, low motility, and abnormal morphology (Lorber, 1992, pp. 169–171). But then a physiologically intact and fertile woman must undergo the medical tests, hormonal injections, laparoscopies, and other procedures necessary for *in vitro* fertilization. It is she who is subjected to medical technology, much more than her husband, and arguably much more than her child-to-be.

In their submission to the Royal Commission on the New Reproductive Technologies, the Canadian Conference of Catholic Bishops listed "The Impact on Women" as one of their seven major concerns (1991, p. 635). Largely because of the articulateness of Canadian women's groups, these bishops have come to understand that the new reproductive technologies "were proceeding without taking into account the physical and psychological consequences to women." The bishops describe in concrete detail a number of experiences which women undergo in using the new technologies. They conclude this section by bringing together their concern about the status of women and their concern for social justice:

> Many women's groups have objected that women are being experimented upon and that their understandable desire to have a child is being exploited. There is a need to listen more closely to what women are saying and to explore more deeply the values of a society which seems ready to spend enormous amounts of money to produce a child at the same time that one in five children in Canada live in poverty because their mothers are poor (1991, p. 635).

VI. SACRAMENTAL THEOLOGY AND MARRIAGE

Richard McCormick has noted that the anthropology of *Donum Vitae* is not dualistic, hence is a post-Vatican II anthropology. But just as *Humanae Vitae* seemed to revert to an earlier theological view, so does the portion of *Donum Vitae* that discusses homologous IVF (1987c, pp. 24–25).

My reading of the *Instruction's* description of the marriage relationship suggests that it has also reverted to an earlier concept of sacramentality.

Various authors have tried to explain why the magisterium regards the sex act and procreation as qualitatively different from all other human acts. Sexual activity seems to be "set in a special category outside normal ways of understanding human activity and exempt from usual ways of doing ethics" (Vacek, 1988, p. 129; Neaves, 1985, p. 13). Cardinal Bernardin supports this singular treatment of sex and procreation, suggesting that there are good reasons for regarding these acts as qualitatively different from all others (1987, p. 25).

But the special position given to these activities leads other authors to ask whether it represents a peculiarly Catholic preoccupation with sex. Is it because sexual rules are made by celibates, who tend to "isolate and overvalue not only the sexual life but also individual sex acts" (Vacek, 1988, p. 129)?

I believe it is plausible that the authors of *Donum Vitae* treat sex and marriage as they do partly because they are influenced by a pre-Vatican II understanding of sacrament. In particular, in their treatment of the principle of inseparability, they are able to slide back and forth between the conjugal act and the marriage relationship, between specific acts of intercourse and the marriage as a whole, because they basically identify the two concepts. And this identification is facilitated by an earlier theology of sacrament.

For centuries the basic concept of sacrament taught in the Catholic Church has been: "An outward sign instituted by Christ to give grace" *(Baltimore Catechism*, 1917, p. 25). More scholarly discussions are found in the works of Catholic theologians, for example, C.C. Martindale, S.J., in his book *The Sacramental System*:

We see then that there exist in the Church certain material transactions, such that they stand as signs of something spiritual, and also, somehow cause and confer and contain what they signify, and that these efficacious signs were in some sense instituted by Christ himself (1928, p. 25).

Thus a sacrament is a physical (material) act or thing which not only symbolizes a spiritual reality, but which actually causes and contains that reality each time it is performed. Apart from gross deviations, this spiritual reality is conferred, or occurs, whether or not the participants are fully present to it or actually experience it. In fact, in the case of infant baptism, the infant need have no awareness whatsoever of the momentous event occurring to it.

The efficacy of the sacraments came to be interpreted by many pastors in a literal and mechanistic way, leading to that interpretation often being taught to the faithful. Nuances such as those employed by Martindale ("*somehow* cause and confer," and "were in *some* sense") were forgotten in much ordinary catechesis. Since Vatican II, the Church has consciously moved away from such literal and mechanistic understandings. Sacramental theology in 1996 emphasizes human experience in its wholeness and totality.

I believe, however, that the authors of *Donum Vitae* have been influenced by an earlier notion of sacrament. They seem to give the act of sexual intercourse the significance or status of such a sacrament within the marriage relationship. Not only is this act the primary symbol or expression of the loving union of the spouses, but it (like a sacrament) contains and accomplishes that union each time it is performed. The reasoning of *Donum Vitae* suggests that this is true regardless of how the spouses experience it, which may vary a great deal from one occasion to another. In the *Instruction*, the specific act of sexual intercourse carries the whole weight of a complex human condition and relationship, marriage.

This perspective enables one to grasp in some way the *Instruction's* identification of the marriage relationship with the conjugal act, as it interchanges one concept with the other. If procreation may not morally be achieved outside of marriage, then, in this interpretation, it may not morally be accomplished apart from the specific act which, in a sacramental sense, *is* the marriage. This sacramental interpretation, while it may represent an outdated theology, is the only way I have been able to overcome the logical incoherence in *Donum Vitae's* two definitions of inseparability and to make sense of *Donum Vitae's* reasoning on the simple case of IVF.

BIBLIOGRAPHY

Abbott, W.M. (ed.): 1965, *Texts Promulgated by the Ecumenical Council (1963–1965)*, America Press, New York.
Baltimore Catechism: 1917, E.M. Lohmann, St. Paul, Minnesota.
Barron, J.: 1987, 'Views of the Vatican document vary from approval to vowed resistance', *New York Times* (March 12), 13.
Bernardin, J.: 1987, 'Science and the creation of life', *Origins* **17**, 21–26.
Brody, B.: 1990, 'Current religious perspectives on the new reproductive techniques', in D.M. Bartels et al. (eds.), *Beyond Baby M*, Humana Press, Clifton, New Jersey, pp. 45–63.
Cahill, L.S.: 1987, 'The Vatican document on bioethics: Two responses', *America* **156**, 247–248.
Canadian Conference of Catholic Bishops: 1991, 'Reproductive technologies and the value of human life', *Catholic International* **2**, 633–637.
Curran, C.E.: 1982, *Moral Theology: A Continuing Journey*, University of Notre Dame Press, Notre Dame, Indiana.
Curran, C.E.: 1988, *Tensions in Moral Theology*, University of Notre Dame Press, Notre Dame, Indiana.
Daniel, W.: 1986, 'In vitro fertilization: Two problem areas', *Australasian Catholic Record* **63**, 21–31.
Donum Vitae: 1987, New York Times (March 11), 10–13.
Ethics Advisory Board: 1978–1979, Transcripts of meetings and public hearings, National Technical Information Service, Springfield, Virginia.
Ethics Advisory Board: 1979, *Report and Conclusions: HEW Support of Research Involving Human In Vitro Fertilization and Embryo Transfer*, U.S. Government Printing Office, Washington, DC.
Falise, M. and Régnier, P.J.: 1987, 'Practice and research at Catholic University of Lille', *Health Progress* (July-August), 63.

Goldman, A.L.: 1987, 'Parts of document stir sharp disagreement from theologians', *New York Times* (March 11), 1 and 13.
Harvey, J.C.: 1989, 'Speculations regarding the history of *Donum Vitae*', *Journal of Medicine and Philosophy* **14**, 481–491.
Jullien, J.: 1987, 'Bioethics: Paradise Lost?', *Health Progress* (July–August), 59–61.
Lehmann, K.: 1987, 'Observations on the interpretation of the *Instruction*', *Health Progress* (July–August), 61–62.
Lorber, J.: 1992, 'Choice, gift, or patriarchal bargain? Women's consent to *in vitro* fertilization in male infertility', in H. Holmes and L. Purdy (eds.), *Feminist Perspectives in Medical Ethics*, Indiana University Press, Bloomington, Indiana.
Martindale, C.C.: 1928, *The Sacramental System*, Macmillan, New York.
Marcus, E.: 1987, 'The meaning of a shock', *Health Progress* (July–August), 56–59.
McCormick, R.A.: 1987a, 'The Vatican document on bioethics: Some unsolicited suggestions', *America* **156**, 24–28 and 39.
McCormick, R.A.: 1987b, 'Document is unpersuasive', *Health Progress* (July–August), 53–55.
McCormick, R.A.: 1987c, 'Begotten, not made', *Notre Dame Magazine* (Autumn), 22–25.
McCormick, R.A.: 1989, *The Critical Calling: Reflections on Moral Dilemmas Since Vatican II*, Georgetown University Press, Washington, DC.
Neaves, W.B. and P.W.: 1985, 'Moral dimensions of *in vitro* fertilization', *Perkins Journal* **39**, 10–23.
Rothman, B.K.: 1989, *Recreating Motherhood: Ideology and Technology in a Patriarchal Society*, W.W. Norton, New York.
Shannon, T.A. and Cahill, L.S.: 1988, *Religion and Artificial Reproduction*, Crossroad Publishing, New York.
Stanworth, M. (ed.): 1987, *Reproductive Technologies: Gender, Motherhood and Medicine*, University of Minnesota Press, Minneapolis, Minnesota.
Tauer, C.A.: 1989, '*In vitro* fertilization: Are there still ethical problems?', *Journal of the Minnesota Academy of Science* **54**(3), 3–7.
Tauer, C.A.: 1990, 'Essential ethical considerations for public policy on assisted reproduction', in D.M. Bartels et al. (eds.), *Beyond Baby M*, Humana Press, Clifton, New Jersey, pp. 65–86.
Vacek, E.V.: 1988, 'Vatican instruction on reproductive technology', *Theological Studies* **49**, 110–131.
Vespieren, P.: 1987, 'Les fécondations artificielles: A propos de l'Instruction romaine sur "le don de la vie" ', *Etudes* **366**, 607–619.
Walters, L.: 1987, 'Ethics and new reproductive technologies: An international review of committee statements', *Hastings Center Report* **17** (Spec. Supp. June), 3–9.

SECTION THREE

MORAL AND SOCIAL REFLECTIONS

DEBORAH D. BLAKE

INFERTILE COUPLES: PSYCHOLOGICAL NEEDS, SOCIAL RESPONSIBILITIES

I. INTRODUCTION

During the past forty years we have witnessed a number of developments in medical technology that assist infertile couples in their efforts to attain biological parenthood. These developments have increased the popular attention given to this most personal issue and to the content of that information (Stone, 1991). The birth of the first *in vitro* baby in 1982 was a technical watershed in reproductive medicine and was accompanied by a gradual shift in the assessment of the psychological aspects of infertility. In the 1950's and early 1960's psychological problems, especially 'women's problems', were presented as a major cause of infertility. Today the fundamental medical focus is on physiological explanations for infertility (Stone, 1991, p. 316). Psychological aspects of infertility have come to be viewed as consequential rather than causative.

Roman Catholic moral assessments of medical procedures to assist infertility have relied heavily upon a physiological-natural law model and only recently have begun to shift the focus to personal, interpersonal and social aspects. Vatican statements on reproductive technologies have employed a version of natural law that relies upon the physical structure of the biological act of intercourse to provide the moral norm used to assess the medical procedures. This emphasis was apparent in Pius XII's statement on artificial insemination and in the Vatican's assessment of *in vitro* fertilization for married couples in 'Respect for Human Life in Our Day' (Congregation for the Doctrine of Faith, 1987). More recent efforts have emphasized relational considerations based upon marital and parental models (e.g. Cahill, 1989; Shannon and Cahill, 1988). Currently, attention is being given to the moral weight and assessment of the consequential psychological needs of infertile couples.

Infertility is experienced as an intensely personal and often private matter. This is reflected in the medical and ethical analyses of reproductive technologies – attention is focused on individuals and couples. The moral discussion is usually pursued as an aspect of personal, sexual ethics rather than as an issue of social ethics. However, Maren Klawiter claims that "[f]ocusing on the individual responsibility for children and individualizing the suffering due to infertility hides the socially-constructed/created problems of child rearing and infertility" (Klawiter, 1990, p. 68). The psychological needs of infertile couples have a social context and a social construction. "One might argue that *in vitro* fertilization and surrogate motherhood can alleviate the pain of infertility. But one might then counter that much of that pain has been socially constructed" (Baruch, 1987, p. 137). The moral assessment of these psychological needs is a complex matter that extends beyond the personal to the social and, consequently, from the arena of personal ethics to social ethics.

Moral consideration of the psychological needs of infertile couples requires a critical assessment of the social and cultural context in which these needs are presented. This is not to dismiss or denigrate the personal difficulties experienced by infertile couples but it is to take seriously the link between our public and private lives. Naomi Scheman argues that "the objects of psychology – emotions, beliefs, intentions, virtues and vices – attach to us singly ... to a piece of ideology ... which ... permeates our social institutions, our lives, and our senses of ourselves...." These are not unalterable but they are deeply infused with particular notions of "liberation, freedom and equality" (Scheman, 1983, 226). Decisions in our personal lives do reflect, and support, ideologies that structure our social reality. Both the dominant, contemporary United States culture and Roman Catholic culture support the personal psychological need/desire to bear children yet they place childbearing within very different constructions of society and social life. The dominant culture of the United States is imbued with a rugged individualism which focuses on the attainment of individually defined goods. Individual freedoms are fundamental; "institutions are as far as possible neutral mechanisms for individuals to attain their separate ends" (Bellah et al., 1991, 10). By contrast, a Roman Catholic perspective views the individual as having an interest in her/his own well-being as well as an interest and moral obligation to participate in achieving the common good of society. Social institutions are means used to achieve the common good. The very personal matter of infertility becomes a socially relevant matter which requires critical moral reflection because personal actions can contribute to the maintenance of a particular social ideology.

In this paper I will argue that the current use of reproductive technologies in the United States, especially the high technology procedures, helps to maintain the dominant social ethos that is fundamentally in conflict with the social ethos presented by Roman Catholic conceptualization of the common good and social justice. This is not an argument about the redistribution of social goods or resources. The central issue is about how personal choices impact social reality and the moral responsibility that one has in her/his personal life for the social life of the community. Roman Catholics who consider the use of reproductive technologies must give serious moral consideration not only to the personal aspects of the situation but to the social impact of their actions understood not merely as direct action but as complicity, the indirect support of particular social arrangements by personal participation and benefit from a particular social arrangement. On a broader scale this argument is an effort to emphasize the importance of both individual and social responsibility for childbearing and child rearing.

II. PSYCHOLOGICAL NEEDS

Infertility is usually defined as the inability to conceive after one year of regular sexual intercourse without the use of contraceptives. In the United States 10% of couples of childbearing age are infertile, 18% of childless couples are infertile, and 50–60% of infertile couples do eventually bear children (Andrews et al., 1991, p. 239). Recent qualitative and quantitative studies have begun to document the psychological aspects of infertility.

A positive correlation between the desire to have a child and degree of stress for an infertile couple has been identified by Barbara Berg and her colleagues (Berg et al., 1991. p. 1075). When asked why they wanted to have a biological child infertile couples indicated:

1. Caring – desire to nurture a child, interest in family life;
2. Self-focus – companionship, to prove parenting ability or adulthood status, to fill a void;
3. Family line/genetic contribution – to carry on family name or genes;

4. Spouse's desire – due to spouse's desire;
5. Biological/social imperative – maternal instinct, adults should do this, reason for existence;
6. Focus on marital bond – expression of marital commitment, to strengthen or salvage marriage (Berg et al., 1991, p. 1076).

Both partners indicated 'caring' with equal frequency. Men were more likely than their spouses to indicate 'spouse's desire' while women indicated 'self-focus' more often (Berg et al., 1991).

A study by Frank Andrews et al. (1991) is representative of the literature on the psychological impact of infertility. In a review of the literature this research team found reports of 'negative psychological, behavioral and social aspects of infertility'. These aspects included:

[1] negative emotions - anxiety, fear, isolation, depression, guilt, frustration, and helplessness;
[2] feeling inadequate, damaged, defective as man or woman, inability to reproduce as evidence of being not quite whole and are a failure;
[3] feeling like a flop, hollow, 'shooting blanks';
[4] a diminished sense of masculinity or femininity, negative body image and self-esteem;
[5] infertile women perceive themselves as less potent than did fertile women (Andrews et al., 1991. p. 239).

This study provided quantitative evidence of the impact of infertility on 'life-as-a-whole' and on 'feelings about marriage and intimacy/sex/romance' that confirmed the conclusions of the qualitative evidence that they found in the earlier literature (Andrews et al., 1991).

The Andrews research team interviewed 157 couples, both husband and wife. The couples were white and middle class. However, none of these couples had yet tried the high technology treatments such as IVF and GIFT. The research data indicated that:

[1] Infertile husbands perceived more home life stress than did their wives ...
[2] Infertile women engaged in more problem-solving and escape-coping than did their husbands ...
[3] Infertile wives attributed more responsibility for the fertility problem to themselves than their husbands did to themselves ...

[4] Infertile wives perceived themselves as having more control over the solution of the infertility problem than did their husbands therefore [they felt] more responsible . . .
[5] Infertile husbands were more satisfied than their wives with the meaning they found in their infertility . . .
[6] Both infertile and presumed fertile women had lower self-esteem and higher sexual dissatisfaction that did their husbands . . .
[7] Women reported experiencing more depression (Andrews et al., 1991, pp. 304–306).

Overall, men and women appeared to differ in their response to infertility; men experienced more overall stress in general quality of life while women experienced stress more directly related to infertility. This work supports earlier claims that infertility can have "far reaching effects on life satisfaction, well-being, and psychological adjustment" (Greil, 1988, p. 179) "Infertility", writes Andrews et al. "Is a crisis that has a psychological toll on individuals and relationships" (Andrews et al., 1991, 298). The impact of infertility on the quality of life is consistently negative and wide ranging.

This emotional crisis has become normalized. Sarah Holbrook notes, ". . . people grow up thinking that conception and giving birth are matters of choice, they experience infertility as an unanticipated crisis and usually react with shock to the discovery of their condition" (Shapiro, 1982 in Holbrook, 1990, p. 333). Popular confirmation of these psychological responses and behaviors is supported, even defined, by popular media. Jennifer Stone observes that articles in popular magazines

problematize 'fertility' and 'infertility' and normalize a set of practices and social relations. These articles encourage individuals to atone for what *Time* calls 'the unforseen consequences of a variety of historical and economic trends' that have caused the 'infertility epidemic.' . . .Putting the locus of responsibility and guilt on individuals who have had the 'bad luck' to live in these times, however, discourages examination of the kinds of social relations that are shaping perceptions of the 'infertility epidemic' and the fear that one might be, without realizing it, scarred, inadequate and deficient. For many, however, the fear is tragically misplaced (Stone, 1991, p. 321).

These psychologically based needs and behaviors are taken for granted, even expected (Spedalle, 1989).

III. SOCIAL CONTEXT AND CONSTRUCTION

Feminists were first to develop a systemic critique of reproductive technology. "[T]he distinctive feature of this critical model ... is the assumption that to critique an emerging technology one must engage in an encompassing cultural critique ..." (Stone, 1991, p. 313). The moral assessment of the personal use of reproductive technologies in the United States must be include a critique of the social and cultural context of the technology. It is the ethos of the *dominant* culture that is under scrutiny here as a context in which the psychological needs of infertile couples are constructed. How are these psychological needs of infertile couples being defined? Whose needs are acknowledged and served by these new reproductive technologies? Finally, what aspects of the dominant culture are sustained by the reproductive technologies?

A. Clients

Those infertile couples who seek medical assistance for infertility are primarily white, middle and upper class, well educated professionals (Stone, 1991; Andrew et al., 1991; Berg, 1991; Klawiter, 1990; Greil, 1988). Yet, the National Centers for Disease Control provide a profile for infertile women that is striking in its difference from those who seek medical assistance; these women are 'older and more likely to be black and have no previous children'. They also tend to have received less than a high school education (Stone, 1991, p. 324). Socio-economic status is a determining factor for access to reproductive technologies. The development of new reproductive technologies has been accompanied by an increase in infertility programs in the United States; in 1990 there were 192 programs (Stone, 1991, p. 323).

Sevgi Aral and Willard Cates point out that the number of doctors treating infertility, as well as those doctors who are board certified in endocrinology/infertility, has increased substantially in the last ten years, at the same time that the birth rate remains low. In addition, changing sexual, marriage and work patterns have 'probably resulted in an increased proportion of women from higher socioeconomic backgrounds among all infertile women'. Competition for customers in the private sector will produce both more attentive and more aggressive care for those who can pay (Gerson, 51).

When asked if reproductive technologies were limited to married, heterosexual couples a research physician in reproductive medicine at a major medical school reported to me, "If you can pay for it you can have anything you want" (Blake, 1992).

B. Marketplace Ethos

Stone links the developments in the medical marketplace with the social construction of the psychological needs of infertile couples. She observes that the

> [s]upply of assisted reproductive service has increased in the last decade, and competition for potential patients is becoming keen. Today there are a number of free-standing independent clinics created with venture capital and dedicated to making money, not just to creating babies. This medical entrepeneurialism, as well as the brutal economic climate within the health care system, has prompted many clinics and hospitals to get more involved in advertising and public relations. The marketing is highly sophisticated and often uses a comprehensive mix of marketing approaches (Cosco, 1988, in Stone, 1991, p. 323).

One element of this marketing is the increase in discussion and advertisement of reproductive developments in popular women's magazines over the recent decades.

The marketing and medicalization of infertility is supported by articles on infertility, direct ads, and advertisers, control over the editorial content of these magazines (Stone, 1991).

> [T]he idea of a 'need' for new technologies to satisfy the 'demand' created by the 'reproductively impaired' was forged through links in discourse generated from multiple, independent institutions and organizations. The need was concretized over 40 years, and there were series of shifts in the meanings associated with infertility (Stone 1991, p. 324).

Mass media are significant contributors to the construction and "normalization" (Spedalle, 1989, p. 98) of the psychological needs of infertile couples, which is to say, the needs of those able to afford the product – white, upper and middle class, well educated, professional couples. "Today the meaning of 'infertility' has been shaped largely by a commercial mass communication system that includes independent mass media industries, public relations, and advertising agencies" (Stone 1991, p. 314).

The broader theoretical framework for the 'marketplace' is described in Lisa Cahill's analysis of the 1988 Office of Technology Assessment's (OTA) report, *Infertility* (Cahill, 1989).

> Th[is] U.S. Government document reflects the liberal and democratic traditions of autonomy and freedom in the quest for happiness as the individual defines it; of tolerance of multiple, co-existing moral and religious definitions of the good and happy life; and of the government's primary role in protecting such freedoms, mediating conflict with the aim of maximizing the liberty of every individual and group. . . . Since free choice is the

dominant value in human relationship, the parameters of marriage, the family and even the parent-child relationship tend to be defined around freely chosen and non-permanent alliance . . . (Cahill, 1989, p. 499).

In the case of reproductive technologies, the expression of freedom is assured or limited according to personal financial resources. The ideal of freedom is tied to the ability, or responsibility, to pay. The responsibility for marriage, family, children, and parenting is individual. There is no sense in which they are tied to the common good. They are tied to individually defined benefit and responsibility. Cahill reiterates,

. . . addressed specifically to the question of public policy in a culture not only pluralistic but aggressively 'liberal' in its value orientations, the OTA report focuses on the freedom of adults to take effective steps to achieve their own reproductive goals, with minimal interference, and with safeguards to ensure that the decisions of some individuals will not impinge unfairly on parallel freedoms of others, now or in the future. Liberty rights are given priority over welfare rights, and rights to use new technologies are generally defined in terms of liberty, not welfare (Cahill, 1989, p. 502).

In this marketplace individual freedom and an individual's uncriticized expression of preferences is aimed at overall maximization of personal satisfaction and avoidance of pain, psychological pain in the case of infertile coupled.

[I]ndividual will or preferential decision and social organization foster[s] a maximal expression and achievement of individual preferences. The generality of the value concept thus expresses the unfettered individualism within a community where the social aim is the maximal achievement of individual achievement (Edel, 1988, pp. 20–21).

To a large degree, the utility of this commodity – reproductive technologies – is constructed by marketplace interests and is available to those who can pay the price. It is this marketplace that contributes to variation in personal reproductive freedom. The degree of personal, high technology, reproductive freedom corresponds to financial well being.

IV. PUBLIC POLICY

A clear message extends beyond the boundaries of the new reproductive technologies – you may freely choose to bear children, if you can pay. While many infertile couples speak of a "right" to bear children the underlying message suggests "privilege". This message becomes a symbolic referent for government policy in the United States since the 1980's regarding childbearing and childrearing. Public policy has exempted government and society from responsibility and placed accountability solely with the parents.

A. Technology as Interpreter

In *Backdoor to Eugenics* Troy Duster argues that research and technologies, especially those that are widely publicized, become interpretive lenses for public perception of social reality and provide the foundation for further public policy that conforms with the original, underlying viewpoint of the research and technology. Duster (1991) demonstrates a link between research projects on the genetic basis for crime and public policy on crime. As with reproductive technologies, there was a marked increase in articles dealing with the genetic basis of crime (Duster, 1990, 93); the interpretive "symbol" was presented to the public. The scientific or medical research model placed the locus of causality with the individual genetic makeup; in the public rendering this translated into individual responsibility. Yet, the very design of the research project yielded a disparity in responsibility across ethnic and socio-economic lines. The studies used twins and incarcerated populations. For the latter, incarceration rates in the U.S. reflect a significant difference in race; African Americans dominate prison population in contrast to their minority status in the national population. No one studied the criminal deviance of those in power. During this same period national "law and order" policy focused on the prosecution and incarceration of individuals to the exclusion of any government effort to examine and respond to social factors, such as poverty and unemployment, which might also have contributed to "criminal behavior". Responsibility for crime in the United States was attributed to individuals with no corresponding consideration of socio-economic conditions as contributing factors.

A "backdoor" to reproductive discrimination has been opened by the new reproductive technologies. Following from the interpretive "symbol" of reproductive technologies the "backdoor" to reproductive discrimination incorporates similar ethnic and socio-economic disparity in the assignment of reproductive rights and responsibilities. It also privatizes infertility, child bearing and child rearing.

B. Reproductive Rights and Responsibilities

This pattern of shifting from social analysis and government interest/ responsibility for social institutions and systems that contribute to the common good to an emphasis of individual freedom and responsibility is appropriate to reproductive and family issues. Suzanne Holland and Karen Peterson (1993) observed that

> issues involving poor women and reproduction are now the subjects of intense debate. In at least two states, California and New Jersey, the reproductive practices of poor women are being scrutinized and blamed for the high costs of welfare. Welfare, it is said, is bankrupting state budgets. Thus, in at least two instances, governors and legislators are targeting poor women and their children as prime redline 'objects' in state budget cutting. In reality, welfare constitutes only 6 percent of the state of California's general fund of $43.8 billion. In some states, legislators are proposing cash incentives for poor women who have the Norplant birth control device implanted and for poor men with vasectomies. This is certainly consistent with previous state and national efforts to make funds available for sterilization of poor women. *It seems that our society would rather eliminate these women's reproductive needs than attend to them* [emphasis mine] (Holland and Peterson, 1993, p. 12).

Socio-economic status is a significant determinant of whose psychological, reproductive needs deserve attention or disregard. Those at the base of the social order are neglected. This disparity is sustained by marketplace strategies and government policies. Consider two public policy issues that correspond to the development of the reproductive technology industry since the 1980's – family leave and restrictions on welfare payments for women who had children while on public assistance.

The emphasis on individual reproductive rights and responsibilities, apart from any analysis of social context and reform of social infrastructure within which one bears and raises a child, is exemplified by welfare reform efforts in California. Governor Pete Wilson proposed "slashing family benefits by as much as 25 percent ... and penaliz[ing] mothers who produce more state-dependent offspring" (Liebert, 1992, p. 216). President Bush promised Wilson that he would do all that he could to help Wilson cut through the federal "web of red tape" (Bush in Liebert, 1992, p. 216). In reply to this policy proposal Mario Cuomo, Governor of New York, asserted that, "So called welfare reform that simply cuts subsistence payments to the poor will not save us. Welfare caseloads have increased dramatically in places like California, New York and New Hampshire – not because of moral weakness but because of political weakness that has produced a failing economy" (Cuomo in Liebert,

1992, p. 216). Cuomo links personal and social responsibility by challenging California's political neglect of the economy. Child bearing and rearing cannot be privatized, they have a social context.

The congressional criticism of the Family Leave Bill and President Bush's veto reaffirmed a position on public policy related to childbearing and family which emphasized individual reproductive and family responsibility while rejecting any degree of social or corporate responsibility; it reinforced the positive correlation of actual expression of personal freedom with socio-economic status. Bush appealed to economic consideration in his veto of this bill, "If our nation is to succeed in an increasingly complex and competitive global marketplace, we must have flexibility in out workplaces ..." (Bush, 1993, p. 12). Bush supported voluntary provision of family leave by employers rather than legislative directive. Marketplace interests defined family interests. Again, family interests were privatized; those who could afford family leave would have access.

In contrast, fellow Republican Senator Christopher Bond testified on behalf of Congressional intervention in the workplace through the passage of legislative policies that support families. Bond testified, "I believe for too long government has ignored the importance of families and in some instances adopted policies more likely to break them up" (Bond, 1993, p. 20). Bond linked governmental responsibility for social policies that support childbearing and family with personal reproductive/family responsibilities. "... we as a society need to make family obligation something we encourage rather than discourage" (Bond, 1993, p. 20).

The link between private life and social life was also made by the U.S. Catholic Bishops. Bishop James Malon testified,

[e]very day the Church sees the need for this legislation. We counsel young couples who are afraid to start a family because they need both incomes to make ends meet. Some women even consider abortion when they could otherwise risk losing their jobs during difficult pregnancies. ...We hear a lot of talk about families – how important they are, how family life should be supported, how family values need to be upheld. What we need now is concrete help, and this Congress can offer modest – but vital – support through this legislation (USCC, 1991, p. 126).

Under recent Republican administrations, family values and the valuing of family extended to those who independently were able financially to support personal families. This kind of public policy action regarding families served to discourage childbearing and families among the financially disadvantaged. Responsibility was placed solely upon individuals with no consideration of the social context of the economic and social institutions that provided positions of reproductive privilege for the economically advantaged and maintained the economic disadvantage of the poor and denied them reproductive privileges. The marketing and popular public perception of the new reproductive technologies provided a persuasive interpretive "symbol" that was consonant with this policy message – you may freely choose the bear children, if you can pay.

V. CRITIQUE: SOCIAL RESPONSIBILITY

The moral problem with the policy message sustained by the current reproductive technologies is not located in the desire to bear children and to form a family but in the bifurcation of personal and social responsibilities for children and families and the negligent disregard of social responsibility by individuals, institutions and government. By focusing on personal 'ability to pay' government and social institutions/systems relinquish responsibility and accountability for the impact of their actions on children and families. By their participation in the 'institution' of reproductive technologies individuals have complicity in the maintenance of a system that focuses on individual self-interest to the neglect of common good, releases government and social institutions from social responsibility for the common good – especially with regard to children and families, and increases the disparity in reproductive privilege between the socio-economically advantaged and disadvantaged.

The issue here is not one of reallocation of funds from the private sector to the public, but of a basic overriding viewpoint and subsequent public policy that stands in contradiction to the tradition of Roman Catholic social teaching. It is on this count that participation in the use of complex reproductive technologies by individuals is morally suspect. A closer examination of Roman Catholic social teaching on the family, common good, and solidarity will begin to lay out the foundations for this claim.

A. Family

Roman Catholic social teaching on the family clearly links social and personal responsibility for the family's well-being. Ernie Cortes argues that "what matters is not just [families'] entitlements, but their relationships to others and the roles of social institutions in their lives" (Cortes, 1991, p. 155). The public policy ethos regarding childbearing and family, supported by the individualistic and material ethos of contemporary reproductive technologies, is one such social institution.

In *Rerum Novarum*, Leo XIII perceived the welfare of families to be an integral part of the common good of society. The family is a 'true society' which precedes the state and has rights prior to the state (Leo XIII, 1891, p. 244).

> ...if the family on entering into association and fellowship, were to experience hindrance instead of help, and were to find their rights attacked instead of being upheld, society would rightly be an object of detestation rather than of desire.
>
> ...if a family finds itself in exceeding distress, utterly deprived of the counsel of friends, and without any prospect of extricating itself, it is right that extreme necessity be met by public aid, *since each family is a part of the commonwealth* [Emphasis mine] (Leo XIII, 1891, p. 244).

Leo XIII presumes the necesssity of both personal and social responsibility to sustain family life; he assumes that diligent labor on the part of the family head-of-household in conjunction with just wages and government mediation in the workplace will ensure family support. Rather than simple entitlements this is a matter of social responsibility, especially on the part of the rich.

The support of families extends throughout the social encyclicals. According to John XXIII, "To [the family] must be given every consideration of an economic, social, cultural and moral nature which will strengthen its stability and facilitate the fulfillment of its specific mission" (John XXIII, 1963, p. 205). The social responsibility of governments for families is made clear in the *Charter on the Rights of the Family*. In this document, writes Ernie Cortes,

> ...John Paul II unequivocally qualifies the preconditions of family life as the entitlements of all people. The Charter outlines a comprehensive list of entitlements deserving protection, including voluntary marriage, control of childbearing, protection of children, family determination of education, support of family in public policy, practice of religion, voluntary association with other families, minimum welfare needs, minimum income, decent housing, and the ability of immigrants to keep their families intact.
>
> John Paul explicitly directs his document to political institutions of national governments and international organizations (Cortes, 1991, p. 160).

In this way the seemingly private realm of the family is intricately linked to the complexity and social responsibilities of social institutions and government (cf. Second Vatican Council, 1965, pp. 288–289; Paul VI, 1971, p. 492).

B. Common Good

The connection between individual, institutional and governmental responsibility can be more clearly understood in light of the Roman Catholic understanding of the common good. The human person has both inalienable dignity and a social nature; society exists for the person, not the person for society. There is a precarious balance here that avoids the extremes of exclusive individualism or totalitarianism. The wellbeing of the person and of the human community are interdependent, one mirrors the other. The good of the whole must "flow back upon persons" (Maritan, 1985, p. 55).

The common good is described in *Gaudium et Spes* as:

> [T]he sum of those conditions of social life which allows groups and their individual members thorough and ready access to their own fulfillment....Every social group must take account of the needs and legitimate aspirations of other groups, and even of the general welfare of the entire human family.
>
> ...Hence, the social order and its development must unceasingly work to the benefit of the human person...(Second Vatican Council, 1985, pp. 263–264).

The desire for children and family finds moral legitimacy in this context wherein these aspirations are met for all, not just for those who are financially and socially privileged. Moreover, a serious moral question is raised for those who, by their current participation in reproductive therapies, help to sustain a public policy ethos that not only neglects but denies the aspirations of those at the base of the social order for family and children.

C. Solidarity

In *Rerum Novarum* Leo XIII recognized and gave special attention to the economic plight of the workers; this provided the foundational base for the development of solidarity with the poor in Roman Catholic social teaching. In subsequent Vatican documents material poverty was linked with the inability to enjoy "basic human rights" (John XXIII, 1961, p. 178), the disregard or the absence of human dignity (John XXIII, 1961, p. 161), and dehumanization (Paul VI, 1971, p. 490). Poverty has come to be understood as marginalization – social, political, cultural, sexual, ethnic, religious and economic. Reproductive marginalization is a further extension of this conceptualization of poverty.

The response of solidarity with the poor begins with a recognition of "the respect due the poor and the special situation they have in society" (Paul VI, 1971, p. 496) and extends to ongoing systemic changes that assist public and private institutions to better the conditions of human life (Second Vatican Council, 1965, p. 266). "Without a renewed education in solidarity", writes Paul VI, "an overwhelming inequality can give rise to an individualism in which each one claims his[/her] own rights without wishing to be answerable for the common good" (Paul VI, 1971, pp. 496–497). In the current climate of reproductive technologies it is this individualism coupled with a liberal materialism that lends momentum to the public policy ethos that supports 'family values' for those who can afford a family and denies reproductive 'privileges' to the economically marginalized.

A fundamental moral concern surrounding the use of complex reproductive technologies is the lack of solidarity with the poor and personal complicity with an unjust structure. Summarizing *Laborem exercens*, Donal Dorr notes that solidarity requires a

dedicated and consistent effort to disentangle oneself from unjust structures, practices, and traditions that keep the poor in poverty; and a serious commitment to building alternatives that will be just and truly human. The reason why it must be done is that we cannot evade responsibility for the injustices that mark our world. Almost everybody has some degree of complicity in these injustices . . .(Dorr, 1983, p. 238).

The fundamental moral issue for reproductive technologies today is not so much the particular structure of the medical procedure or the desire to have children but the unreflective detachment of individual/partner decision-making from the social, political and economic context. It is the assumption of autonomous liberty devoid of recognition and responsibility for the common good.

VI. CONCLUSION

The psychological needs of infertile couples, as described in the research literature, focus on individual needs and interests. The Roman Catholic assumption of the inalienable dignity of the person as well as the person's essential social nature requires that serious moral weight be given to the experienced reality of the individual but this cannot occur outside of the social, communal context. The individualism and materialism of the dominant culture in the United States predisposes its citizens to overlook the social and communal aspects of the experience of infertility. The liberal emphasis on individual freedom is interpreted as a right of non-interference and freedom from social obligation. This is far from the Christian understanding of freedom which entails responsibility for self and for one's community. Moreover, this liberal ethos suggests that individual fulfillment is detached from the good of the community; nothing could be further from the Roman Catholic tradition of the common good.

The moral analysis of the use of reproductive technologies for infertile couples needs to extend beyond the usual arena of 'personal ethics' to incorporate 'social ethics'. This suggests that our personal choices and actions do have consequences for the life of the community. More critically, it suggests that our personal benefit may be linked to the disenfranchisement of others in the larger social network; if it is, then we share moral responsibility for the injustices of that system. The dominant liberal culture in the United States rejects such a viewpoint; Roman Catholic social teaching supports it.

BIBLIOGRAPHY

Andrews, F.M., Abbey A. and Halman L.J.: 1991, 'Stress from infertility, marriage factors, and subjective well-being of wives and husbands', *Journal of Health and Social Behavior* **32**(3), 238–255.

Aral, S. and Cates, W., Jr.: 1983, 'The increasing concern with infertility, why now?', *JAMA* **250**(17), 2327–2331.

Baruch, E.H.: 1987, 'A womb of his own', *Women and Health* **13**(1/2), 135–139.

Batterman, R.: 1985, 'A comprehensive approach to treating infertility', *Health and Social Work* **10**(1), 46–54.

Bellah, R.N., Madsen, R., Sullivan, W.M., Swindler A. and Tipton S.: 1991, *The Good Society*, Alfred A. Knopf, New York.

Berg, B.J., Wilson, J.F. and Weingartner, P.J.: 1991, 'Psychological sequelae of infertility treatment: The role of gender and sex role identification', *Social Science and Medicine* **33**(9), 1071–1080.

Betzig, L.: 1989, 'Causes of conjugal dissolution: A cross-cultural study', *Current Anthropology* **30**(5), 654–676.

Blake, D.D.: 1992, Notes from a conversation with a research physician in reproductive medicine at a major U.S. medical school.

Bond, C.S.: 1993, 'Floor Debate on President's Veto of Family and Medical Leave Act', *Congressional Digest* **72**(1), 20–24.

Budde, M.L.: 1992, *The Two Churches: Catholicism and Capitalism in the World-System*, Duke University Press, Durham.
Bush, G.: 1993, 'The President's veto message', *Congressional Digest* **72**(1), 12, 32.
Cahill, L.S.: 1989, 'Moral traditions, ethical language, and reproductive technologies', *The Journal of Medicine and Philosophy* **14**, 497–522.
Congregation for the Doctrine of Faith: 1987, 'Instruction on respect for human life in its origin and on the dignity of procreation: Replies to questions of the day', *Origins* **16**, 697–711.
Cortes, E.: 1991, 'Reflections on the Catholic tradition of family rights', in J.A. Coleman, S.J. (ed.), *One Hundred Years of Catholic Social Thought*, Orbis Books, Maryknoll, NY.
Daly, K.: 1990, 'Infertility resolution and adoption readiness', *The Journal of Contemporary Human Services* **71**(8), 483–492.
Dorr, D.: 1983, *Option for the Poor*, Orbis Books, Maryknoll, NY.
Edel, A.: 1988, 'The concept of values and its travels in twentieth-century America', in M.G. Murphey and I. Berg (eds.) *Values and Value Theory in Twentieth-Century America*, Temple University Press, Philadelphia.
Final Document of the Third General Conference of the Latin American Episcopate (Pueblo, Mexico): 1979, in J. Eagleson and P. Scharper (eds.), *Pueblo and Beyond*, Orbis Books, Maryknoll.
Frank, D. and Vogel, M.: 1988, *The Baby Makers*, Prometheus Books, New York.
Gerson, D.: 1989, 'Infertility and the construction of despair', *Socialist Review* **19**(3), 45–64.
Greil, A.L., Leitko, T.A. and Porter. K.L.: 1988, 'Infertility: His and hers', *Gender and Society* **2**(2), 172–199.
Holbrook, S.M.: 1990, 'Adoption, infertility, and the new reproductive technologies: Problems and prospects for social work and welfare policy', *Social Work* **35**(4), 333–337.
Holland, S. and Peterson, K.: 1993, 'The health care Titanic: Women and children first', *Second Opinion* **18**(3), 11–29.
John XXIII: 1963, *Pacem in Terris* in J. Gremillion (ed.): 1976, *The Gospel of Peace and Justice*, Orbis Books, Maryknoll, NY.
John XXII: 1961, *Mater et Magistra* in J. Gremillion (ed.): 1976, *The Gospel of Peace and Justice*, Orbis Books, Maryknoll, NY.
John Paul II: 1988, 'Sollicitudo rei socialis', *Origins*, **17**(38).
John Paul II: 1983, *Charter on the Rights of the Family*, Vatican Polyglot Press, Vatican City.
Klawiter, M.: 1990, 'Using Arendt and Heidegger to consider feminist thinking on women and reproductive/infertility technologies', *Hypatia* **5**(3), 65–89.
Kraft, A.D., Polombo, J., Mitchell, D., Dean, C., Meyers, S. and Schmidt, A.W.: 1980, 'The psychological dimensions of infertility', *American Journal of Orthopsychiatry* **50**, 618–627.
Liebert, L.: 1992, 'Pete Wilson and welfare', *California Journal* **23**(4), 216.
Lorber, Judith: 1987, '*In vitro* fertilization and gender politics', *Women and Health* **13**(1/2), 117–133.
Maritain, J.: 1985, *The Person and the Common Good*, University of Notre Dame Press, Notre Dame, IN.

Paul VI: 1971, *Octagesima Adveniens* in J.Gremillion (ed.): 1976, *The Gospel of Peace and Justice*, Orbis Books, Maryknoll, NY.

Rapp, R.: 1987, 'Moral pioneers: Women, men and fetuses on the frontier of reproductive technology', *Women and Health* **13**(1/2), 101–116.

Scheman, N.: 1983, 'Individualism and the objects of psychology', in S. Harding and M.B. Hintikka (eds.), *Discovering Reality: Feminist Perpsectives in Epistemology, Metaphysics, Methodology and Philosophy of Science*, D. Reidel Publishing Company, Boston.

Second Vatican Council: 1985, *Gaudium et Spes* in J. Gremillion (ed.): 1976, *The Gospel of Peace and Justice*, Orbis Books, Maryknoll, NY.

Shannon, T.A. and Cahill, L.S.: 1988, *Religion and Artificial Reproduction*, Crossroad Press, New York.

Shucker, E.: 1987, 'Psychological effects of new reproductive technologies', *Women and Health* **13**(1/2), 141–145.

Stone, J.: 1991, 'Contextualizing biogenetic and reproductive technologies', *Critical Studies in Mass Communication* **8**, 309–332.

USCC (United States Catholic Conference of Bishops): 1991, 'Testimony on Family Leave Bill', *Congressional Digest* **70**(4), 124–126.

Valentine, D.D.: 1986, 'Psychological impact of infertility: Identifying issues and needs', *Social Work in Health Care* **11**(4), 61–69.

PATRICIA BEATTIE JUNG

WHAT PRICE FERTILITY?

I. INTRODUCTION

Most ethical evaluations of *in vitro* fertilization (IVF) focus on two important moral considerations. They tend to analyze this whole cluster of new reproductive technologies in terms of their impact (1) on the inviolable sanctity and dignity of every human life and (2) on the relationship between sexuality, marriage and procreation. Questions such as these dominate the moral literature on IVF: what is the nature of the moral relationship between sexuality, marriage and procreation? Is it a necessary relation? In what sense? Is a zygote a human person? Or, as the American Fertility Society names its governing norm, is each person involved in these procedures "integrally and adequately considered"? Traditional arguments revolve around both the validity of these two principles and their applicability to IVF.

These are important considerations, thoroughly addressed elsewhere in this volume. I wish in this essay to focus on an additional set of moral questions, not altogether absent from discussions of this issue among moral theologians, but certainly not prominent. I wish to ask: who, if any one, should have access to these technologies? Should IVF have a high priority among our public health care allocations in order to make it more accessible? Should Catholic health care institutions play a prophetic, if not automatically counter-cultural role regarding such allocation decisions?

Such queries did not of course originate with me. In his review of "The Religious Response to Reproductive Technology," Arthur L. Greil (1989, p. 12) notes that for some time a few ethicists have argued that it is "immoral to spend time and resources on extraordinary means of promoting births when attention should be devoted to preventing unwanted births and improving the health of all infants". Of course that doesn't necessarily invalidate Greil's conclusion that these arguments have not been a part of "the most important religious objections" to the new reproductive technologies. Such a narrow reading of what is at stake in regard to IVF is typical of the field. Indeed in his review of several international committee statements on ethical issues surrounding new reproductive technologies, Leroy Walters (1987, pp. 3–9) found no discussion of these issues.

II. EXPANDING OUR ETHICAL FOCUS

In this essay I will press for the adoption of a particular (in this case restrictive) policy regarding public funding for and institutional support of IVF services. The success of such an endeavor rests in part on convincing people that the consideration of such matters (regardless of where one lands in the deliberative process) is of great importance in the first place. The tasks of moral theology include not only problem-solving but problem-setting as well. The latter task involves the evaluation of how we see and define moral problems. It is my contention that moral theologians must enlarge the lens through which IVF is typically examined. There are at least three reasons which justify such an expansion of our ethical focus. The first two rationales have to do with the biomedical and socio-political contexts in which we actually make judgements about such matters. The third is derived from and related to more traditional concerns.

A. *Scientific Context*

First our scientific context recommends this expanded focus. We live in a world exploding with technological possibilities. Thus it is not possible to make an all-encompassing, categorical (whether the ruling is restrictive or not) judgement about IVF. Actually IVF is not a single procedure but is a growing, ever-changing family of reproductive techniques which includes not only IVF and ET (embryo transfer) but also ZIFT (zygote intra-Fallopian transfer), GIFT (gamete intra-Fallopian transfer) and PZD (partial zona drilling for the sake of improving implantation rates) among others.

It is possible that one or more of these procedures, or some related techniques to be developed in the near future, may not violate the two traditional concerns more typical of moral arguments about IVF. If one or more of these techniques proved to be morally acceptable on these traditional grounds, a third group of previously overshadowed concerns should surface. A full ethical evaluation of IVF requires that we ask: (1) to whom such (otherwise at least hypothetically permissible) procedures should be accessible, and (2) what constitutes a prudent and just expenditure of our health care resources in regard to this new family of reproductive technologies?

B. Socio-Political Context

Second our social context demands that we assume the wisdom of such expenditures. Few in the United States would deny that our health care system is in trouble. Soaring costs, growing inaccessability and increasingly uneven patterns of distribution all contribute to the escalating sense of crisis here. Although the particulars vary significantly, people around the globe face a fundamental choice. We have the resources (1) to make hypothetically unlimited medical care available to some, or (2) to make minimally decent but clearly limited health care available to all. We do not have the resources to do both. In this sense the world in which we live – a world of limited resources bursting with technological possibilites – forces a choice upon us.

Each course of action entails rationing, although the patterns are quite different. In this sense the rationing of health care is inescapable. The important moral question is not whether we should ration health care, but what norms should govern our allocation of health care.

This growing socio-political recognitioon of and concern about competing demands for limited health care resources is especially relevant to new reproductive technologies for two reasons. First, the demand for all kinds of infertility treatment is already high and on the rise.[1] Depending upon how you define infertility, at least 10% of all heterosexual couples experience it and as many as 20% of all heterosexual couples experience significant contraceptive difficulty. In 1990 alone over a million new patients sought treatment for infertility. How ought we weigh their claims against other competeing demands for health care and other social services, such as public education? Competing demands will require that we make hard choices.

The answer of course is that we have already made our choice. In one way we already do ration IVF. This brings us to the second reason for attending to the way reproductive technologies in general, and IVF in particular, are allocated. Many who desire this treatment and who could potentially benefit from it do not have access to it. This fact is for them inevitable and already painful.

Generally speaking only the wealthy who are able to pay out of their own pockets for the technology are assured access to it. This "marketplace" approach to the rationing of IVF means that only those who are members of higher income groups can realistically afford IVF. Usually it is not available to those of the middle class. Many members of this group are underinsured, or have otherwise fairly comprehensive coverage but policies which exclude these particular services.

Presently most insurance companies exclude IVF. This is why in 1987 after intensive lobbying Massachussetts was among the first states to enact laws which mandate that insurance companies pay for all infertility treatments, including IVF. Perehaps as many as nine states now require such coverage. According to Lori B. Andrews (1989, p. 387) "in Maryland, a law was passed requiring that insurers that cover other pregnancy-related services must pay for IVF as well, if certain conditions are met". This of course does not impact the lower classes. Few, if any, members of this group have access to any (comprehensive or otherwise) insurance coverage, and there is at present in the United States little, and then quite restricted, public funding for such procedures.

Since there is no valid basis for assuming that people of lower socioeconomic classes would not make good parents, a question arises about our present "marketplace" approach to rationing IVF. Is it just? Is IVF part of that level of health care (variously called minimal, basic and/or adequate) to which we ought to guarentee universal access? In countries which do have national health care systems, at least some reproductive technologies are so subsidized, usually artificial insemination by donor (AID). There is a growing movement in many of these places to include IVF among procedures so covered. What is society's obligation to infertile couples?

C. Extending Traditional Concerns

Including in our deliberations about IVF an assessment of how, if at all, it ought to be distributed is important for other reasons. These reasons are in fact extensions of traditional concerns. Concerns about distributive justice and IVF are related in principle to more typically traditional evaluations. For example, the conviction that every person has a right to adequate health care and that access to such basic care must be provided to all is rooted in and is an expression of belief in the sanctity of human life and the dignity of all persons. These principles do not determine what constitutes a basic level of health care, but they do ground a concern with such matters.

Frequently judgements about whether fertiltiy is a medical need or merely a desire hinge in part on how one conceives of the relationship between sexuality, marriage and procreation. In his pamphlet, "*In Vitro* Fertilization", Paul T. Jersild (1986, p. 7) describes questions about IVF's considerable demand on our medical resources and the relative priority of such procedures as growing from profound theological convictions about the nature of parenthood. He argues that when "the responsibilty of parents before God is centered not in child-bearing but in child-rearing" enabling parents to care for their children may take priority over assisting conception. (We shall return to this point later in this essay.)

Suffice it to say that there are several good reasons for expanding the focus of our evaluations of IVF. Whether, and how, such a family of technologies might be justly distributed deserves careful moral attention. Present day scientific and socio-political realities, as well as traditional moral concerns, call for such an analysis. Sketched in what follows are several aspects of such an evaluation.

Though not a comprehensive outline of all these issues, important dimensions of the way commitments to the good stewardship and just distribution of our resources are considered. My analysis begins with a discussion of how this problem might be fruitfully framed. I take it to be axiomatic that while persons have no right to give birth, they may legitimately claim under ordinary circumstances to have a positive right to basic health care. It is within this framework that this problem is most sharply defined. Does the positive right to adequate health care entitle persons to controceptive assistance – specifically to IVF?

III. FRAMING THE PROBLEM

In its instruction on "Respect for Human Life in Its Origins and on the Dignity of Procreation: Replies to Certain Questions of the Day" (*Donum Vitae*), the Congregation for the Doctrine of the Faith (CDF) (1987, p. 34) demonstrated that no true and proper positive "right to a child" can flow from the desire for a child. To treat children as the object of such a right, even if one's motive is to relieve the suffering of the infertile, would be to violate their dignity and nature as gifts.

Whether Catholic or not, most biomedical ethicists would concur with this conclusion. Few would argue that there is a positive right to give birth or to have a child. Some might argue that there is *prima facie* a negative right to noninterference in one's reproductive decisions, but this does not entitle one to financial assistance. Those who contend that IVF ought to be made accessible to all usually do so on other grounds – on the ground that people have a right to adequate health care.

It is possible of course to argue against making new reproductive technologies widely available (regardless of a couple's ability to pay) on the premise that there is no such right – not even to minimally decent health care. But it is not necessary (or legitimate, in my opinion) to make the case on that ground. Conceivably all parties to these discussions may concur that respect for persons requires that we make available to all a certain "basic" level of health care. But that is where the consensus is likely to end.

Those who contend that infertile people are entitled to new reproductive technologies, including IVF services, perceive inequities in fertility rates to be both unfortunate and unfair. Those who are infertile are seen as entitled to community resources in their quest for a remedy of this condition. This conclusion does not rest solely on the basis of a claim about a positive right to health care. Several additional premises, about which there is much contention, are central to the argument.

This essay highlights two of these. First, in order to justify such an expenditure of resources it must be established that infertility is an illness or disease which deserves medical treatment (see section four). Second, it must be established that the provision of IVF is essential to adequate health care. This entails demonstrating, among other things, that the personal and social costs of the treatment are proportional to its benefits, and to its canons of justice (see sections five and six).

IV. WHAT TO LABEL INFERTILITY?

It is not clear in what sense infertility is an illness or disease. Human reproductive potential, especially when explored from the female's point of view, is naturally experienced as mercurial, episodic and restricted to just one season of a woman's life. Many couples eagerly anticipate the onset of menopause, with the accompanying demise of fertility. Most persons who are infertile are not otherwise ill. Infertility is clearly only a problem when the couple has a strong desire to have a first or another child.

This issue is crucial to our analysis. Some would argue that fertility is most properly described as a wish or desire, not a medical need at all. Others contend that though not life-threatening, infertility is a grave health problem because of the emotional suffering associated with it. I wish to argue that infertility is a condition for which both the physiological causes and symptoms are appropriately subject to medical response (though not every such treatment is thereby justified.)

The desire to have children is both natural and socially constructed. It would be terribly naive (if not deceptive) to deny that those various conditions we identify as having a legitimate claim for medical care are at least *in part* socially constructed. Both human needs and desires are in this sense quite elastic and interrelated. This does not contradict the CDF's contention that "on the part of spouses, the desire for a child is natural: it expresses the vocation to fatherhood and motherhood inscribed in conjugal love" (1987, p. 33). It merely helps us to recognize that such interpersonal desires have not only physiological bases but are always experienced in particular cultural contexts.

Few in our society seriously debate the positive value of having children. As a whole they are perceived as contributing to the survival of the species, to the enhancement of society and to the pleasure and advantageous self-development of their parents. But disagreements about the relative importance of having children have generated debates among us about how best to weigh claims for reproductive assistance against other medical and social claims. Some want to argue that having children is more than valuable: it is fundamental to being human or to personal self-fulfillment, at least or especially for women. This of course moves far beyond recognizing the value of having children.

Whether one agrees with it or not, it cannot be denied that such a view of parenthood would inevitably feed what Daniel Callahan might call the "technological imperative" in reproductive matters, and that it looks suspiciously like an ideological artifact of that very technology. In the words of Linda Williams, "the society in which a woman lives (may) create a market for IVF by placing so many important meanings on fertility that to be infertile indeed becomes an unbearable problem (1990, p. 229). She encountered women who reported in their interviews that they felt their very identity as women challenged by their infertility. Infertility carried for them the stigma of sexual inadequacy and failure.

In a world where having children is portrayed as the sole criterion for assessing a women's life or for assessing the fruitfulness of a marriage, it is not hard to imagine why many experience infertility as unbearable.

Many, myself included, believe that having children is profoundly meaningful. Nevertheless when this script for human fulfillment is portrayed as an ideal exhaustive of a person's (usually a woman's) potential, it is unduly stifling, even oppressive. In their instruction the CDF makes it clear that while having children may be a great blessing to one's marriage, it is not requisite for the fulfillment of conjugal love.

I would add here that theologians of the church might do even more to clarify for average communicants the exact nature of the relationship between sexuality, marriage and procreation in the light of the widespread misunderstanding that the significance of having children may foster in couples an obsessive attitude toward having children. Such desperateness may render them extremely vulnerable to exploitation and unresponsive to ecclesial instruction regarding such matters as well.

Of course one need not adopt such delusions about parenthood in order to recognize that those who are infertile may suffer from a condition appropriately subject to health care. A rejection of exaggerated claims about parenthood results only in the unraveling of the belief that infertility is a *grave* disorder. It does not threaten its basic identification as a disease. Instead it challenges claims that such treatments should have a *high* priority among health care allocations which tragically must increasingly be limited in order to be more justly distributed.

V. LIMITED HEALTH CARE RESOURCES

Presently in the United States we spend 12% of our GNP or approximately $800 billion annually on health care. Many people believe this allocation to be more than reasonable. Societies must place limits on their health care expenditures, balancing them against other needs for housing, education, employment etc. Some would argue that we are overspending in this regard and that we must cut back this line item in our national budget. (Only 6% of the GNP is paid to the National Health Service in the United Kingdom.) Virtually all agree that no additional resources should be allocated to health care in this country. At a minimum we must cap our spending.

Of course in one sense we have already "capped" our spending, by excluding many from basic health care coverage. An estimated 35.4 million people in the United States are without health insurance, and 60 million more Americans are underinsured. For a variety of reasons many of these men, women and children fall through the national "safety net" and have access to no health care at all.[2]

I take it to be axiomatic however that this pattern of "rationing" is unjust. Presently our resources are inequitably distributed, and justice demands that we increase access to basic health care services for a substantial number of people. Since virtually all agree we ought not achieve this goal by increasing our overall health-related expenditures, improving access to health care will necessitate at a minimum that our system become far more streamlined and efficient. Some measure of cost-containment will be achieved through the adoption of measures designed to increase the cost-effectiveness of our delivery system.

However, the truth about our situation is that a significant degree of cost-containment will be achieved only when we begin to make some tough decisions about our health care priorities. Both the good stewardship of our financial resources and the demands of justice require that we recognize the limits of our health care resources. As a rule it is good to avoid needless sorrow and/or to remedy disappointment whenever possible. But it is also generally recognized that suffering is not to be avoided at all costs.

A. Maximizing the Good We Would Do

Concretely this means we must ask what measure of our resources we should allocate to the treatment of infertility, and specifically whether it is justifiable to provide IVF services when the basic health care needs of many others languish and worsen. Cost-benefit analyses should inform (though not necessarily determine) such allocation decisions. This will prove to be a complicated business.

First we must ask whether the burdens of IVF are in proportion to the very real but limited value of having a first or another child. The emotional cost associated with IVF are difficult to measure: it is frequently associated with anxiety and depression, and almost always takes the patient-couple on an emotionally wrenching, roller coaster ride. There are some health risks, both fetal and maternal to factor into the equation. Success rates are notoriously variable: overall it is less than 10%, though it is probably not unreasonable to anticipate that the efficacy of these procedures could double in the near future.[3] Costs (excluding these associated with travel, lodging and lost wages) are estimated to range from $5,000–$9,000 per menstrual cycle. In order to offer the average IVF couple a 50% chance of a live birth, society would need to be prepared to spend approximately $38,000–$50,000 per couple. Furthermore, a shift toward making these technologies universally available, even if accompanied by appropriate "consumer" cost-sharing mechanisms, is likely to intensify the already escalating demand for these services.

IVF is expensive and often futile, and the ratio between its benefits and burdens poor. Yet most infertile couples choose and would likely continue to avail themselves of it. The grief, indeed bereavement, of those who wish to have a or another child is profound, and like all those who find themselves stunned by loss it is difficult ro see beyond that pain.[4] Because our resources are limited however it is inappropriate to continue to think of this as solely a matter of individual choice.

The relative priority of the needs of infertile couples in relation to other demands on the health care system ought to be discerned in a communal process. I believe that in such a deliberative process, the extent of the other, presently unmet needs for primary, preventive and curative health care would dwarf the need for IVF. In a widely participatory process public funding and institutional support for IVF would (and I believe should) be given a low priority in comparison to other presently unmet health care needs.

However it is important to recognize that a decision not to provide such clinical services ought not to be made lightly, even when communally based in a fully participatory, comparative analysis. Even though IVF is costly and of limited effectiveness, its nonprovision is not good. This is so because it is not obviously disproportional in itself. Like the occasional decision to go to war, this judgement may be justified but it ought not be celebrated. It is justified because IVF had a poor cost-benefit ratio *compared* to other imaginable health care expenditures. However such a policy is best viewed as a tragic necessity. Furthermore its rationing can only be justified if the resources thereby "saved" are used to bolster all people's access to basic, more beneficial services.

VI. THE COMPLEXITY OF DOING JUSTICE

Cost-benefit analyses alone do not guarantee that our health care resources will be distributed justly. It can be reasonably argued that special, preferential treatment ought to be given to some among us on the basis of the principle of fair opportunity, even when such treatments have comparatively poor cost-benefit ratios. In the light of this principle one might argue that infertile couples are in an important sense "disable" and therefore ought to be given preferential treatment – in the form of comparatively cost-ineffective treatments – even at the expense of meeting other basic health needs. It would be morally naive not to expect that the demands of justice may on occasion *conflict* with those of beneficence. Initially arguments for equal opportunity seem compelling and straight forward in their implications. A longer look at them proves just how complicated doing justice is.

A. *Redefining Infertility*

If infertile couples are "disabled" because they cannot give birth; parents without adequate family health care coverage are similarly "disabled" because they cannot meet the basic health care needs of their children. Having children is not simply about childbearing; it is about child rearing as well. If birthing is not to be reduced to the mere "production" of a child, then the tasks and goals of parenting – of human fertility – must be understood as encompassing not only conception, gestation and birth but all the demands for nourishment, care and generativity that follow birth as well.

Those who wish to enable parenthood through IVF on the grounds that all people deserve a fair opportunity to parent need to wrestle not only with statistics about infertility, but with these as well. Presently the United States ranks 19th in the world with regard to infant mortality. We rank 17th with regard to highly cost-effective preventive health care measures for children, such as polio immunization programs. In 1985 we spent less on elementary and secondary education than 14 other nations.

It is not at all clear that the provision of new reproductive technologies is at the heart of our obligation to give people a fair opportunity to parent. Indeed in this particular case the demands of justice as expressed in the principle of fair opportunity may *reinforce* rather than conflict with our obligation to maximize the good that we do by giving IVF a low priority.

B. Other Justice Concerns

I believe a case can be made that it is only just to ration IVF if at least some of the resources so "saved" are channeled back into more cost-effective infertility treatments, like screening programs equally available to all designed to prevent infertility. This seems only fair in light of the fact that the rationing of IVF asks those infertile couples suitable for it to shoulder more than their fair share of the community-wide responsibility for meeting the other health care needs of all.

I would argue further that the rationing of IVF can only be justified if it applies to all, not merely those without the ability to pay. For the same reasons that we ought not to publicly fund these procedures for those who cannot privately afford them, we should not put our institutional resources at their service or into their development. Other needs should have priority not only for our tax dollars, but for our professional talents, and hospital space and laboratory equipment. It has long been understood that social justice can sometimes only be purchased at the expense of certain individual liberties. Justice does not come cheaply. It demands that the wealthy infertile and those health care professionals with keen clinical and research interests in this field pay a high price – again, more than their fair share – in order that all have access to basic health care. This sacrifice ought to somehow be communally recognized.

VII. THE GLOBALIZATION OF ADOPTION

Interestingly, among those who seek out IVF services are couples with other children (either adopted or biological), as well as the childless. They include many more than those who wish to experience pregnancy, or desire to have a genetically related child. Indeed according to the structured interviews of Linda Williams (1990, p. 232), "the fact that they would have a genetic bond with an IVF baby seemed less important than the fact that an IVF baby would be a second child." This would suggest that our faltering domestic "supply" of children available for adoption, and regulations that rule out older persons as suitable adoptive parents, may be as much a factor behind the desire for IVF services as is clinical infertility itself.

In many countries boys are valued over girls, and adoption is virtually unthinkable because of the cultural importance placed on blood ties. As a consequence there are millions of abandoned and otherwise homeless children, especially girls, abroad. No doubt there is much to consider when evaluating the worldwide search for adoptable children on the part of many U.S. couples. Still, it is worth careful consideration. There is more than one way to have children.

NOTES

[1] A variety of cultural factors and medical interventions contribute to this growing epidemic. For example, economic realities and career aspirations press women to delay childbearing until they are thirtysomething, and the availability of highly effective contraceptive techniques enable many of them to succeed at it. The generally unrecognized risk in this script for women is that women grow naturally less fertile with age. Many young women who wish eventually to bear children know that the onset of menopause will signal that they have run out of time. What they don't seem to understand is that the so-called "biological clock" will not only run out of time; it gradually runs down. Female infertility naturally increases with age.

Other factors contribute to this epidemic as well. The diminishing significance of fidelity in the sexual lifestyles of many had resulted in a sharp rise in the rates of sexually transmitted diseases (STDs). Many STDs impair fertility, especially Gonorrhea, Herpes Simplex II, and Chlamydia. Even among those whose infertility is congenital a strictly natural, biological lottery may not be to blame. Some women who were born in the 50s and 60s are infertile today because their mothers took a drug (DES) during their generation to prevent miscarriage. It had an unfortunate consequence: the malformation of female reproductive systems.

To attempt on the basis of this kind of information to distinguish the "worthy" or "innocent" victim of infertility from the "irresponsible" sufferer is to miss the point of this entire rehearsal of the social factors relevant to infertility. To note that we have some measure of personal responsibility for our own sense of health does not justify neglect of our neighbor. To allocate health care on the basis of merit is to operate with a highly individualistic sense of both what contributes to our health and of who is morally responsible for it.

[2] The growing inaccessibility of health care in the U.S. can be attributed to a number of factors. Two of the most important are (1) eroding employer-based insurance coverage and (2) unreasonable eligibility requirements for Medicaid.

[3] A standard method of calculating success needs to be established. Presently clinics vary as to whether they base their estimates on a successful attempt at fertilization, a biochemical pregnancy, a clinical pregnancy or a live delivered infant.

[4] As is the case with other losses, their crisis is spiritual as well as interpersonal. The Gracious Giver of Life is experienced as the Arbitrary Withholder of Life; the fruitfulness of their love is experienced as withered.

BIBLIOGRAPHY

Andrews, L.B.: 1989, 'Alternative modes of reproduction', in S. Cohen and N. Taub (eds.), *Reproductive Laws for the 1990's* Humana Pr., Clifton, New Jersey, pp. 361–403.

Congregation for the Doctrine of the Faith: 1987, 'Respect for human life in its origins and on the dignity of procreation: Replies to certain questions of the day', United States Catholic Conference, Washington, D.C., pp. 1–40.

Greil, A.L.: 1989, 'The religious response to reproductive technologies', *The Christian Century* **106**(1), 11–14.

Jersild, P.T.: 1986, '*In vitro* fertilization', Lutheran Church of America/Division for Ministry in North America, New York, pp. 1–12.

Walters, L.: 1987, 'Ethics and new reproductive technologies: An international review of committee statements', *The Hastings Center Report* **17**, 3–9.

Williams, L.: 1990, 'Wanting children badly: A study of Canadian women seeking IVF and their husbands', *Issues in Reproductive and Genetic Engineering: A Journal of International Feminist Analysis* **3**(3), 229–234.

KEVIN WM. WILDES, S.J.

IN VITRO FERTILIZATION: SECULAR MORAL AUTHORITY, BIOMEDICINE, AND THE ROLE OF THE STATE

I. INTRODUCTION

One of the most overlooked sections of *Donum Vitae* is, perhaps, the most important chapter of the document. The third chapter of *Donum Vitae* is entitled "Moral and Civil Law." The chapter argues that there are natural moral values, concerning marriage and family, that must be protected by those who govern civil society. This chapter of *Donum Vitae*, with its legislative recommendations, is logically consistent with the natural law tradition that has been so influential in the shaping of Roman Catholic moral theology. The assumption of natural law theology is that a common morality can be discovered by natural reason. Since the moral law can be known by reason, society ought to be shaped by it. Since the role of the civil government is to promote the common good and civil order it is the duty of civil government to enforce morality. *Donum Vitae* reasons that general principles can be known and applied universally and since such principles about marriage and the family can be discovered by natural reason, it is a duty of civil governments to regulate new reproductive technologies.

The view of the state expressed in *Donum Vitae* is not unique to Roman Catholicism. In his commentary on English law Sir William Blackstone notes that the law of nature, the moral law, is one of the foundations of the civil law. The civil law builds upon the moral duties one has to God, neighbor, and the self and it fulfills a *remedial* function when people violate the natural laws or fail in their moral duties (Blackstone, pp. 42–55). This view of the law was aptly articulated in this century by Lord Patrick Devlin who understood that the state should "compel a man to act for his own good" (Devlin, p. 136). This view of the state and civil law assumes that society can know what the good for each person is and that the state has the moral authority to enforce it.

Donum Vitae expresses a view of morality, the state, and civil society that reflects an ancient hope of the West. The hope of discovering a common moral law was expressed by the Roman jurists Gaius and Justinian who held that the *jus naturale* was known to all animals and that the *jus gentium*, the law of nations, embodies what reason commands of any rational agent (See, Justinian, Book I. 2.). This hope shaped a general view of morality and the state that was influential well into this century. The third chapter of *Donum Vitae* repeats this hope. But it does so at a time when this vision of morality seems more dream than realistic hope. Contemporary moral controversies, such as those in biomedicine, often turn on disputes over differing expectations of the moral good, moral duty, and the good of human life. These fundamental disputes led to skepticism about the hope that a common morality for marriage and the family can be discovered. The fundamental problem for a natural law methodology is that there are many different accounts of common morality and the content of the natural law.

This essay seeks to explore the role of the state, in a secular society, in enforcing morality. The first section will examine the appeal to the natural law. The second section will turn to the epistemological issues presented by moral language and moral pluralism. Then the essay will see how these problems play out in bioethics. Finally, the essay will examine the appropriate role of the state.

II. NATURAL LAW

An historical review of the natural law tradition and the appeal to nature reveals the weakness of this framework as a basis for moral argument in a secular, morally pluralistic world. It is clear that there is no natural law tradition but *there are natural law traditions*. And therein lies part of the problem.

The Stoic philosophers are some of the first to appeal to nature as a basis for determining human conduct. The Stoic view centers on the assumption that the whole universe is governed by laws which both exhibit rationality and which can be discovered by human reason. Human beings, following the rest of the universe, also have essential nature which is law governed. Determining what is the *essential* nature is a prerequisite for determining the moral constraints.

One approach to distinguish those aspects of human life that are essential from those that are non-essential is to appeal to the ends or purposes of human life. One finds this type of teleological model in Aristotle and Thomas Aquinas. Aristotle thought that there was a defining end to human life, eudaimonia, which is tied to the flourishing of human nature. For this to occur one needed to be schooled in certain virtues. Aquinas brings together the Stoic principle that we ought to "follow nature" with an Aristotelian view of nature as a teleological system. For Thomas natural moral law is part of divine reason that is accessible to human intelligence (ST, q 91, a. 1–3). It is not to be confused with the biological or physical order which have their own laws. The precepts of the natural moral law take the form of something to be done. That is, they outline the basic human goods that ought to be protected and pursued. The pursuit of these goods is tied to Aquinas's view of human essence and ends. The basic goods are pursued so as to effect the human. The goods are ordered so that one may achieve the end of human life (happiness or blessedness). In turn the goods give moral constraints and directions for how one ought to act. That is, they direct us in what we should pursue and what cannot be done. We find here a very particular view of human nature that is tied to assumptions about the divine nature of all creation, particularly the human, and the divine ordering of creation.

One of the shifts that takes place in natural law thought is a move away from the theological and teleological frameworks of the natural law. By the seventeenth century the appeal to nature was used by the sciences as a way to talk about the natural world. The emphasis was no longer on the human participation in the divine law. Instead the emphasis was on regular, measurable realities of physical nature. Teleology was abandoned for different forms of mechanization. The appeal to nature no longer had, necessarily, moral dimensions.

These shifts away from religious, teleological framework to mechanistic views of the world are reflected in the natural law thinking of the seventeenth and eighteenth century. Following the Protestant Reformation, Grotius and others moved away from the theological and teleological metaphysics used by Aquinas. They argued that the fundamental nature of the human being was the possession of reason. They were not concerned with developing an account of basic human goods to support a certain teleology. Rather the natural law could be understood as what reason discovers. Of course this position leads to further questions about the nature of reason.

This brief recounting is not meant to be an exhaustive overview of the natural law. The point of the overview is to illustrate a profound conceptual issue for any appeal to human nature. Any appeal to nature as the basis of moral constraints depends on how nature is understood for the constraints that will be developed. Outside the context of any particular moral framework there will be numerous ways to understand nature. Absent a common held view of human nature, we will not be able to develop content-full moral constraints.

At first glance many people will think that an appeal to nature is sufficient for moral guidance. Yet, as one investigates this appeal it becomes apparent that to understand the meaning of an appeal to nature one must situate the term in context of moral language. Outside of a particular context of a moral language terms such as 'nature' can take on multiple meanings so as to become meaningless. One can recall the thirty year debate in Roman Catholicism about the moral evaluation of birth control. There are those who evaluate the physical act and see artificial contraception as illicit while those who argue for a person-centered ethic see the act as morally neutral and put the evaluative emphasis on the intention of the agent. Yet both arguments appeal to nature.

The problem then is clear. If people agree that an appeal to nature is paramount in our moral analysis it will depend on which understanding of nature is brought to bear on the issue. Appeals to nature are embedded within the context of a moral world view. Different appeals will often be incommensurable with one another even though they all claim to appeal to nature.

Even if we could resolve the first fundamental question about the nature of 'nature' we are left with a second, perhaps, more difficult issue: Why should nature be morally normative at all? That is, unless one begins moral analysis by viewing nature, however defined, as morally normative, then there is no reason as to why one should think nature as morally normative. We are caught in a vicious circle. This objection revisits the concerns raised by Hume's analysis of the is-ought distinction and G.E. Moore's concern about the naturalistic fallacy. Hume argued that simply because we can describe how something is does not mean that we can deduce an ought from the is. Even if we could establish a common understanding of human nature, we would still face the question of why nature should be normative morally. If one sees nature as the outcome of random chance there will be no reason to view nature as normative. Indeed, one may well hold the view that the natural moral imperative is to give rational, human control of the designing process.

This circle is not simply a philosopher's game. In a morally pluralistic world the assumptions with which one begins moral argument are significant. That is, unless men and women share the same assumptions (about nature) then they will not follow arguments that hold that nature gives us moral constraints.

III. THE FRAGMENTATION OF MORAL LANGUAGE

The dilemma of contemporary moral language is really twofold. First, confronted by a pluralism of moralities, rather than a single moral narrative, moral terms will take on different meanings in different moral perspectives. Institutions like marriage and family are here very instructive since there are more and more a variety of understandings as to what constitutes "family" and "marriage". The conceptual difficulty is that in general, secular discussions moral terms often lack the shared meanings and premises to establish a normative criteria as to what constitutes a proper "family" and "marriage. Contemporary moral controversies over reproduction and same sex marriages exemplify this dilemma. There will be different accounts of proper sexual and reproductive behavior in a secular, morally pluralistic society. Yet there is no way to choose between these accounts. In a secular state, with many different moral visions, moral terms and institutions can lose their argumentative power and become foundations of babel.

Even though there is no shared moral narrative to define moral institutions like marriage and family they are, nonetheless, often invoked to justify coercive state interventions. However, confusion develops because, while people may use the same words, the moral terms and phrases often have very different meanings. For some "marriage" may mean a sanctified union of man and woman while for other it may mean a commitment of two same sex partners.

The example of Captain Cook's encounter with the Hawaiians is instructive in several very important respects. First, it presents the epistemological dilemma created by moral pluralism. Cook's party was shocked because the Hawaiians thought it permissible, in some circumstances, to live together without the benefit of marriage. However, the Hawaiians were upset by the fact that European men and women ate meals together. This was a practice which, for them, was taboo (Cook, pp. 91–95). Each group had different views about what constituted morally appropriate behavior. Yet, there was no common criteria by which they could determine which was the correct set of moral nomos.

The clash of moral viewpoints makes clear the epistemological challenge of determining which moral standards *ought* to bind men and women. For a moral argument to be rationally convincing people must share the premises on which the argument is built. Without such basic agreement the arguments will be rationally unconvincing. MacIntyre captures this dilemma in the title of his book *Whose Justice? Which Rationality?*. The moral conclusions one reaches depend on the assumptions one makes about moral rationality and the moral world.

The Cook example is instructive is a second respect in that when asked by the Europeans to explain the taboo the Hawaiians could not. They knew it was taboo for men and women to eat together, but they were unable to explain *why* this was the case. The post-modern condition is not only marked by moral pluralism but by a condition in which moral language has been separated from its foundations. There is the danger that a community may come to no longer understand its own traditions (See, Keenan, this volume).

The particularity of moral language is an affront to Western moral philosophy which long believed that moral content could be discovered by reason. The modern philosophical hope has been that reason could discover a content-full secular morality which all men and women share in virtue of their nature as reasonable agents. The conceptual difficulty lies in justifying the basis for particular view of moral rationality. Without particular moral commitments moral arguments remain vacuous. For example, the first principle of natural law, (Do good and avoid evil) (Aquinas, I-II, 94, a.2), needs content if it is to guide practical reason. One needs to specify a ranking of goods and harms if one is to know what is the good to be done and the evil to be avoided. However, absent a canonical ranking of values which is shared by, or binding upon, all a content-full moral argument cannot be developed (Engelhardt, 1996 chapter two).

Bioethics has sought to avoid the conceptual difficulties encountered by theoretical models of moral reason by appealing to casuistry and the model of middle level principles as alternative strategies.

IV. BIOETHICS

Bioethics confronts a scylla and charybdis. Without a content and particular view of moral reason one cannot hope to resolve the moral controversies of biomedicine. However, with a content one looses universality by becoming parochial and the solutions to moral dilemmas will only satisfy those who share the initial premises. Unless one can resolve this dilemma it will be impossible for the state to impose particular moral practices in biomedicine short of using coercion. Bioethics has tried to navigate these waters by adopting different strategies.

One of the best known strategies is the appeal to middle level principles without foundation in any particular theory (Beauchamp and Childress). The appeal to middle level principles, however, is fraught with problems. First, there is the difficulty in determining what the principles mean. One suspects, for example, that when Beauchamp and Childress claim to have agreed on their four middle level principles one is witnessing a slight of hand. For a preference utilitarian 'autonomy' will be understood as the liberty to achieve certain goods. For a preference utilitarian, it will mean, in part, respect for individual choices. For a deontologist 'autonomy' is concerned not with the pursuit of heteronomous desires but with acting in accord with the demands of reason imposed by the moral law. Childress, even though he does not endorse a Kantian deontology, relies upon right making and wrong making criteria which are independent of consequences. So while utilitarians and deontologists can both speak of 'autonomy', the words function in radically different languages and will, in addition, have quite different meanings. The preference utilitarian may argue that out of a respect for marriage and family one must honor the preferences of individual agents. In this view, marriage and family could justify the practice of same sex marriages when chosen by competent individuals. This understanding of the marriage and family would be used to oppose the use of state authority to forbid same sex marriages. In stark contrast, for a deontologist, the marriage and family could be interpreted to mean a reverence for sexuality such that the deontologist would support the use of state authority to stop same sex marriages and regulate reproductive technologies.

A second difficulty with the middle level principles approach is that it is never clear how the principles should relate to one another. Each of the principles is conceived of as prime facie binding. If the principles are all of equal weight, it is not clear how we are to decide, in a conflict situation, which principle to follow. A final difficulty facing this approach is that there is no clear justification as to why this particular set of principles should be canonical. One might wish to include their principles which have been omitted from their list. There is no agreed upon starting point which allows us to select the principles we think should be followed or to give the principles content.

To resolve these dilemmas, the middle level principle model would have to be recast in the context of a theoretical account which would define the meaning, relations, and justification of the principles. In Western philosophy principles have traditionally been a part of a comprehensive structure which begins with some first principle(s) and moves to secondary (middle level) principles (see, Aquinas, I-II, q. 94, aa. 1-6). Seeking to avoid the difficulties of an overabundance of theoretical accounts, Beauchamp and Childress have attempted to excise the secondary principles from any type of comprehensive structure. They hope in this way to avoid the challenge of providing foundations. However, shorn of theoretical and contextual moorings the principles become ambiguous in meaning and impotent in resolving moral dilemmas.

A second attempt to offset the difficulties of moral theory have been an effort to revive moral casuistry (Jonsen and Toulmin). Jonsen and Toulmin offer an historical account of casuistry and allege that an appeal to casuistry can resolve moral controversies in our pluralist, post-modern secular world. However, they never develop an account of how secular casuistry should function. Traditional casuistry was built upon the analysis of particular cases and their resolutions within a concrete content-full moral tradition. Within that tradition certain cases and their resolution could be regarded as paradigmatic for moral dilemmas. The difficulty with a secular casuistry is that there is no way to decide which cases are to function as the paradigm cases because again, like the appeal to middle level principles, casuistry, shorn of a moral viewpoint cannot select the paradigm cases to make the machinery run. Even if a secular casuistry could develop a set of paradigm cases outside of any particular theoretical and cultural framework or content, there is still no non-arbitrary way to describe the moral dilemmas in need of resolution and choose which case should be the model to resolve it because one's description of a case will depend on one's moral point of view.

Since casuistry, according to Jonsen and Toulmin, builds moral principles from paradigm cases then any content-full moral principles, such as those concerning marriage and family, will depend on the paradigm cases selected to resolve a controversy. In the controversy over reproductive medicine paradigm cases will give very different meanings to the institutions of marriage and family. Indeed *Donum Vitae* has a very clear understanding of the models of marriage and family that should govern the uses of reproductive medicine. The crucial difficulty, for a secular casuistry, is that there is no non-arbitrary way to decide which paradigm to apply (Wildes, 1993).

V. THE STATE: LIBERTARIAN BY DEFAULT

The difficulty for secular bioethics is that physicians, patients, citizens, and biomedical scientists frequently meet as moral strangers: without a common starting point or shared moral vision. There is no starting point, in theory, principles, or casuistry, which is not particular or arbitrary in some way. Each starting point must presuppose some moral content as well as a particular notion of reason to develop resolutions to moral dilemmas. Yet the content is purchased at the price of universality. The proposed solutions in bioethics to the moral fragmentation of the post-modern age face the same difficulties of moral fragmentation as do more general moral theories.

If one cannot appeal to a particular concept of God or to a particular understanding of moral rationality or to a particular understanding of nature, in order to ground bioethics, there is only one source left: the authority of moral agents. If one cannot discover an authoritative moral vision to ground moral judgments then one must appeal to persons as the source of moral authority. When men and women meet outside a particular understanding of morality, they have only each other to whom they can appeal in order to resolve moral disputes and in order to frame the fabric of moral interactions. It is for this reason that one finds the salience in the post-modern worldof such practices as free and informed consent, the free market, and limited democracy (i.e. governments that recognize robust rights to privacy, areas where content-full moral views of the majority cannot be imposed on those in the minority). Absent agreement on external moral standards the only moral standards possible are those derived from the agreement of persons as moral agents.[1]

If one is interested in resolving issues peaceably without recourse to force and with moral authority that can be justified in a general secular moral terms, then moral authority can only be derived from the agreement of persons as moral agents. In such circumstances one cannot discover who is *a* moral authority, but only who is *in* moral authority. One cannot rely on reason to discover moral authority, but must rely on the moral will to create moral authority. The necessary condition of mutual respect (the non-use of others without their consent) is not grounded in a value given to autonomy, liberty, or persons but is integral to the project of controversy resolution when God is not heard by all in the same way and when reason has not succeed in establishing a general, canonical content-full moral vision. A morality for moral stangers requires one not only to refrain from using others without their consent, but also to acknowledge them as agents who can agree to or refuse to collaborate. In the secular context of moral strangers one acknowledges, for example, that others may have very different notions of family and marriage or, perhaps, no recognizable notion at all.

Appeal to mutual respect allows us to understand in general secular terms (those which do not depend on a particular moral vision) when force and coercion can be justified in general secular terms. Moral strangers may use each other only when they act with commonly conveyed moral authority. Those who use others without consent loose a commonly justified basis for protest when they are met with punitive or defensive force. Limited democracies draw upon the morality of mutual self respect to provide protection from and punishment for the unconsented-to use of persons (e.g., murder, rape, burglary) as well as to insure the enforcement of the contracts. Bioethics has developed a number of procedural mechanisms, (e.g., advanced directives, informed consent, institutional review boards), which are designed to help mena dnwomen of diverse moral viewpoints collaborate. These are procedural solutions which enable health care institutions and practitioners to navigate tje plurality of moral commitments. While recognizing the rights to privacy they can create, through common consent, endeavors such as a basic health care system (limited solidarity).

Here rights to privacy are not celebrated becasue of any positive value assigned to such rights. Rather they mark out the limits of plausible moral authority of the state to intervene in the peaceable consensual actions of individuals. It is within the enclave that rights to privacy mark out that differing interpretations of sanctity of life can flourish.

In the West the state has often been advanced as the guardian of society's moral culture (e.g., Lord Devlin). From the betrothal of Church

SECULAR MORAL AUTHORITY, AND THE ROLE OF THE STATE   191

and state by Constantine through contemporary legal decisions that state has been understood by many as the protector of public moral culture. However, the inability of reason to discover a content-full canonical morality raises the as to which moral culture the state is to protect and enforce. In view of the contemporary fragmentation of moral views, and the foundational epistemological difficulty of discovering a common moral view, the role of the secular state, as protector of public moral culture, is brought into question.

There is a role for the secular state insofar as it provides a framework in which moral strangers, those with differing and diverse moral points of view, can meet and cooperate peacefully. While the moral authority of the state is limited it is nevertheless crucial to a secular society. The state becomes central to protecting the rights and exchanges of moral agents. It must punish those who unjustly take from others by force or deception. It must enforce agreements that have been freely made. The state becomes an agent for allocating commonly held public resources. However, given the plurality of moral values, and understandings of moral rationality, the state will not have moral authority to enforce a substantive moral vision of sanctity of life since it cannot be known what the vision should be.

The practical implications of this account of the moral authority of the state is illustrated in the practice of reproductive medicine in limited democracies. Given the exclaves of privacy in the secular state, men and women can establish limits to their medical care. Such limits are announced in advanced directives to determine which medical care is to be provided and which is to be withheld. At the same time professional groups and health care institutions can establish guidelines for the types of medical care they will offer and how they will deploy their medical resources and expertise. Institutional and professional guidelines may specify the withdrawal of treatment when it is judged to be futile or even stipulate criteria for the exclusion from treatment based on considerations of quality, length of life, and cost. Since there is no canonical, content-full view of morality which supports the authority of the secular state, individuals, professionals, and institutions will have to articulate their view of appropriate medical action. The state cannot, legitimately, enforce a content-full view of the moral obligations of medicine.

The state will be called upon in biomedicine, and in all fields, to enforce a morality of procedure rather than a morality of content. Absent a common moral narrative, moral terms such as sanctity of life will be incoherent in general secular discussion. Confronted with moral pluralism the authority of the state to enforce a particular morality evanesces. The role of the state is to ensure the free, peaceable exchanges of its citizens. It functions to enforce agreements, protect unconsented-to violations of privacy, and to distribute commonly held resources. The enforcement of content-full morality becomes the work of particular communities rather than the work of the state. Under the protection of the enclaves of privacy many understandings of marriage and family can flourish. The post-modern condition poses a challenge for communities with particular moral visions to participate in public discussions of the issues of reproductive medicine. At the same time such communities cannot look to the state to do more than can be justified in general, secular terms.

The recognition and understanding of the post-modern predicament leads to a recasting of the way the moral authority of the state is understood: the secular state should be morally neutral with respect to content-full understandings of the good life including content-full understandings of family and marriage. The state can, in general secualr terms, have the authority to protect its citizens from unconsented-to-encroachments, to enforce agreements, and to distribute commonly held resources.

Yet there is a reluctance to embrace the limited moral authority of the state. The reluctance stems, in aprt, from the remaining bits and shards, to which people cling, of a once dominant and coherent morality. These pieces of the past, common words used by differnet moral communities, often give rise to the hope of a common morality as a basis for the state's moral authority. Jonsen and Toulmin, Beauchamp and Childress take common moral language and common morality as their point of departure. Indeed Jonsen and Toulmin take the work of the National Commission for the Protection of Human Subjects as evidence of a common morality to which to found their casuistry (National Commission). However, moral terms, while use din common, often have meanings which can only be understood within a particular moral context. Such common moral terms, like sanctity of life and menschenwurde, are the remnant of an earlier, coherent moral view which is no longer canonical for a secular, morally pluralistic society. In such a society there will be no one content-full view that is taken as canonical by all. Secualr moral discourse becomes evacuated of all content and the hope to give the state greater moral authority seems misplaced. However, in discourse with moral friends, moral language will have content and meaning which cannot be articulated in general secular terms alone.

CONCLUSIONS

The epistemological issues made evident in moral pluralism raise profound questions about the extent of the state's moral authority. These issues provide real conceptual and practical challenges to the view expressed in chapter three of *Donum Vitae*. Such a challenge does not mean, however, that the Church, or any community, is voiceless in "the public square." The Church can enter into discussion with the different elements of the wider *society*. It can raise issues about the use of resources, the treatment of women, the understanding of children that are often involved in the use of these technologies. Perhaps, most importantly, it can give witness, within the community of the Church itself, to the values it proclaims.

NOTE

[1] The author of this paper holds that objective standards of morality do exist. The conceptual difficulty for secular moral philosophy is epistemological. There is no 'view from nowhere' to enable us to know, by reason alone, which are the correct standards.

BIBLIOGRAPHY

Aquinas, T.: 1948, *Summa Theologica*, Christian Classics, Westminster, Maryland.
Beauchamp, T. and Childress, J.: 1989, *Principle of Biomedical Ethics* (Third Edition), Oxford University Press, New York.
Blackstone, W.: 1969, *Blackstone's Commentaries*, St. George Tucker (ed.), August M. Kelley, New York, Vol. V.
Cook, J.: 1893, *Captain Cook's Journal 1768–71*, Capt. W.S.L. Wharton (ed.), Elliot Stock, London.
Devlin, P.: 1965, *The Enforcement of Morals*, Oxford University Press, London.
Englehardt, H.T.: 1996, *The Foundations of Bioethics*, second edition, Oxford University Press, New York.
Jonsen, A. and Toulmin, S.: 1988, *The Abuse of Casuistry*, University of California Press, Berkeley.
Justinian: 1922, *Institutes of Justinian*, T. Sandars (trans.), Greenwood Press, Westport, CT.

Keenan, 1996, 'Moral horizons in health care: Reproductive technologies and catholic identity', *Infertility: A Crossroad of Faith and Technology*, Dordrecht, Kluwer Academic Publishers, pp. 53–71.

Lyotard, Jean-Francois: 1984, *The Postmodern Condition*, Manchester University Press, Manchester.

MacIntyre, A.: 1981, *After Virtue*, University of Notre Dame Press, Notre Dame.

MacIntyre, A.: 1988, *Whose Justice? Which Rationality?*, University of Notre Dame Press, Notre Dame.

Wildes, K.: 1993, 'The priesthood of bioethics and the return of casuistry', *The Journal of Medicine and Philosophy* **18**, 33–49

SECTION FOUR

CONTINUING CONVERSATION

RON HAMEL

EPILOGUE

Humpty Dumpty sat on a wall,
Humpty Dumpty had a great fall,
All the kings horses and all the king's men,
Couldn't put Humpty together again.

Is this the fate of *Donum Vitae's* moral analysis of homologous *in vitro* fertilization (IVF) and embryo transfer (ET)? At the conclusion of these essays, almost all of which seriously critique the Instruction's treatment of the issue, one is left with the impression of an argument, fallen, and irreparably shattered. Does the cumulative effect of the critiques leave the Instruction's case in pieces? And have the contributors to this volume made *their* case that the Instruction's moral analysis is inadequate? While I will leave these judgments to the reader, I do wish to draw together the common as well as varied concerns of the contributors in order to provide a better sense of the cumulative impact of the individual assessments of *Donum Vitae's* prohibition of IVF in the "simple case," that is, using the gametes of husband and wife. These concerns tend to fall into three categories: methodology, argumentation, and assumptions.

The contributors' objections to the methodology employed by *Donum Vitae* in its moral assessment of *in vitro* fertilization and embryo transfer between husband and wife are four: (1) inadequate use of its natural law methodology; (2) narrowness in its range of considerations; (3) misunderstanding of the moral object, and (4) inconsistency in its sacramental theology. First, *Donum Vitae* claims to employ a natural law methodology and to base its conclusions on natural law reasoning. As Tauer, in particular, observes such reasoning depends on reflection upon human experience as lived and as studied by various disciplines. Given this starting point, one would expect broad consultation on such fundamental matters as marriage, sexuality, procreation, parenthood, and infertility with the broad range of individuals who could shed light on these out of their own experience; with couples who have resorted to *in vitro* fertilization and clinicians and others who have worked with these couples and performed the procedures; and, of course, with a broad range of experts from various pertinent disciples, including theologians and ethicists. However, it appears, that in preparing the document, the authors did not consult widely. If they did, it is not reflected in the document. And if they didn't, it seriously compromises the adequacy and force of the Instruction. It is difficult to take seriously a moral argument which takes little or no account of human experience and of those in the best position to interpret that experience. Given this, the document's force is likely to be external (based on authority), rather than internal (based on the merits of the argument).

Hence, what should have been, at least in part, an inductive methodology ended up being a deductive methodology which applied a universal principle, namely, the inseparability of the unitive and procreative aspects of sexual intercourse, based on an understanding of the nature of marriage and the conjugal act. This principle itself, however, is assumed and not argued. Wider consultation and increased attention to the experience of married couples would likely have provided a richer and more adequate context for a discussion of the principle and the conclusion to which it points.

A second objection to the methodology employed in the Instruction is the narrowness of its range of considerations. *Donum Vitae's* point of departure in its moral analysis of reproductive technologies consists primarily in the principles of personal sexual ethics and ignores the principles of social ethics. Because of the methodology chosen, the range of considerations is narrowed largely to whether or not homologous *in vitro* fertilization is consistent with the inseparability principle. Consequently, the Instruction neglects an opportunity to surface and assess a number of assumptions underlying the development, advocacy, and use of IVF and other reproductive technologies, and fails to address very serious ethical issues of a social nature – such as who has access to these procedures, use of public funds to pay for them, and the like. The failure here, as Blake suggests, is not only a failure to address social issues intimately related to reproductive technologies, but also a failure to draw upon the Church's rich social justice tradition. This is, in addition, a lost opportunity to assume a prophetic stance by underscoring the social dimension of reproductive decisions. In casting their moral analysis in the categories of personal ethics, the Vatican has simply reinforced the cultural tendency to view reproductive decisions as matters of individual choice. But as Blake argues, one of the central issues in reproductive decisions is how personal choices impact society and the moral responsibility that individuals have for the social life of the community, for the common good.

Yet a third methodological difficulty is the Instruction's understanding of the "moral object." Keenan argues that the authors of the Instruction have both mistaken the moral object and have defined the moral object in a way that is inconsistent with the best of the tradition, in particular, the legacy of Thomas Aquinas. Keenan maintains that Aquinas never defined the object as an "act in its physicality but rather in its agency." The moral object was the proximate content of one's intention. It was the object of intention that was judged as either virtuous or vicious, and not the physical act as good or evil. Keenan believes that *Donum Vitae* employs both methods at various times. On the one hand, it affirms that biological parenting cannot be intentionally separated from marriage. On the other hand, it argues against most reproductive technologies, including homologous *in vitro* fertilization, on the basis of a moral assessment of a physical act. As Keenan observes, in using this manualist logic, the Instruction ends up drawing lines about what actions may and may not be performed instead of holding up what values should be protected, promoted, and expressed. If Keenan is correct, and there is good reason to believe that he is, then there is a serious flaw in the substance of the Instruction's argument which, as Keenan suggests, casts doubt on the validity of its conclusions.

There is a final methodological difficulty to be noted – an inconsistency in the document's anthropology and sacramentality. While the Instruction is generally not dualistic in its anthropology, it reverts to a pre-Vatican II anthropology in the section dealing with homologous IVF. The anthropology there not only tends to be dualistic but also focuses on the physical structure of the act. Tauer believes *Donum Vitae* also reverts to an earlier sacramentality in the way it discusses the conjugal act and marriage. The reason the authors of *Donum Vitae* are able to move back and forth between specific acts of intercourse and the marriage as a whole is because they identify the two. This is facilitated by a notion of sacrament that views sacrament as a physical (material) act or thing which not only symbolizes a spiritual reality, but which actually causes and contains that reality each time it is performed. Not only is the conjugal act the primary symbol or expression of the loving union of the spouses, but it contains and accomplishes that union each time it is performed. This perspective according to Tauer enables one to grasp in some way the Instruction's identification of the marriage relationship with the conjugal act, as it interchanges one concept with the other. If procreation may not morally be achieved outside of marriage, then, in this interpretation, it may not morally be accomplished apart from the specific act which, in a sacramental sense, *is* the marriage.

A second recurrent theme in these essays is the inadequacy of or a disagreement with *Donum Vitae's* argumentation. There are a number of different issues raised along these lines. Tauer notes several. The Instruction, she observes, often substitutes assertion for reasoned argument. That is to say, the Instruction often states a principle and follows it immediately with a conclusion without actually explaining how it reached that conclusion. There are times, in fact, when the same principle could just as easily have been used to reach the opposite conclusion, for example, that homologous IVF can be morally justified. In addition to this lack of reasoned argument, Tauer believes the document also is flawed by what she calls a "series of non-sequiturs." One example she gives is the affirmation of a non-dualistic view of the person (p. 134) followed in the next sentence by a conclusion based on the physical as decisive.

There are other "logical ambiguities." Perhaps the most serious which Tauer notes has to do with the inseparability argument. Does the inseparability of the unitive and procreative refer to the conjugal act or to the marriage relationship? Tauer maintains that at various points in the document, it refers to both. While the magisterium has consistently held that the first definition is correct, it does, on occasion, employ the second. The first it employs to prohibit homologous *in vitro* fertilization. But it is a logical impossibility, Tauer argues, "to maintain both definitions of the inseparability principle simultaneously". Furthermore, Tauer contends (and Porter as well) that there is a problem in the very use of the inseparability principle in the moral assessment of IVF. Inseparability in the narrow sense only applies if there is an act of intercourse, but in IVF there is no act of intercourse. Hence, in the narrow sense, inseparability entails no conclusion about the moral acceptability of IVF. Tauer suspects that the authors must have realized this since they slip into the broad sense in condemning IVF. This, she maintains, is a fallacy of equivocation.

Several authors challenge *Donum Vitae's* assertion that homologous *in vitro* fertilization is an unloving act. This is a claim made without support. In fact, the opposite could be demonstrated. On a more theoretical level, Tauer suggests that there is no reason why the use of medical techniques to achieve fertilization should be inconsistent with a parental love that is procreative. The use of IVF could well be a supreme example of love, of mutual self-giving. Ponticas and Fagan would agree with this, based on their experience with couples seeking IVF. They maintain that these couples demonstrate a great commitment to have a child together and have invested a great deal personally, emotionally, and financially in doing so. Not only does the Instruction make a claim without empirical evidence, which evidence would likely show it to be wrong, the claim itself is offensive to and demeaning of couples who pursue IVF. It is judgmental in attributing to them sinister motives and is likely false in most cases. The charge that children conceived as the result of medical techniques cannot be the fruit of parents' love is not only a non-sequitur, but is also an insidious and biased assertion. It is curious that *Donum Vitae* is so ready to assume that a conception resulting from IVF is an unloving act, yet says nothing about the numerous conceptions that result from acts of sexual intercourse which are unloving acts. There is something deeply troubling about this.

Yet another inadequacy in the document's argumentation is the judgment that homologous *in vitro* fertilization is immoral because it "is deprived of its proper perfection." But again, the Instruction resorts to assertions rather than argument. Why is it the case that actions which lack proper perfection are immoral? Or is it only certain actions which lack proper perfection that are immoral? We are not told. IVF is not the most ideal manner in which to conceive new life, but does that fact make it morally wrong? The authors of the document do not help us in understanding this.

Carlson also raises questions about the logic and consistency of the argumentation in *Donum Vitae*. He wonders, in particular, why the Instruction condemns IVF and embryo transfer while it approves procedures (without specifying which) that "assist the conjugal act either in order to facilitate its performance or in order to enable it to achieve its objective once it has been normally performed" (p. 107). Or why theologians who support the conclusions of the Instruction justify TOTS (tubal ovum transfer with sperm) and GIFT (gamete intrafallopian transfer), but not IVF. These procedures are viewed as assisting and supplementing the conjugal act and not replacing it. They modify the gametes' "route and sequence" in the effort to assist fertilization. The interventions of physicians and technicians are understood as acts *subordinated to* the conjugal act, with the whole carefully sequenced toward the common goal of procreation. But if TOTS and GIFT can be subordinated to the couple's conjugal act, why can't the same be true of IVF in similarly relevant circumstances? And if the route and sequence of gametes can be morally adjusted, why not the site of their coming together? What is the moral significance attached to the fallopian tube as opposed to a lab dish? Further, it is not clear to Carlson why conception in a laboratory must of necessity be viewed as treating the embryo as an object.

Finally, Wildes contends that there are serious difficulties with *Donum Vitae's* argument that state authorities should regulate the development and use of reproductive technologies. This, he argues, is an inappropriate request. The Instruction asserts that there are natural values concerning marriage and family, which can be discovered by natural reason, that must be protected by those whose responsibility is to promote the common good and civil order. Wildes, however, challenges the assumption that there is a common morality that can be discovered by reason. In the post-modern predicament, a common morality cannot be discovered or agreed to by all. Consequently, one is left with particular moralities which include different perceptions of the moral justifiability of reproductive technologies. Given the fragmentation of moral views and the epistemological difficulty of discovering a common morality, it is impossible to justify state enforcement of a particular morality. The state's role is to remain neutral in the face of disparate moralities and instead to ensure free, peaceable exchanges among citizens, the appropriate locus for differences to be resolved.

In sum, a good deal of the discussion in these pages is that the methodology and argumentation of *Donum Vitae* are seriously flawed. Some would probably hold that they are fatally flawed, so that at least some of the Instruction's conclusions are called into question. But poor or bad methodology and argumentation are not the only concerns of the contributors to this volume. A number of authors seek to challenge assumptions associated with reproductive technologies as they are reflected in *Donum Vitae* as well as in the culture.

One of the most prevalent and influential of these assumptions is the significance of parenthood as a measure of personal fulfillment and achievement. Alleviation of these psychological reactions is sometimes considered a justification for reproductive technologies. On the surface, the psychological impact of infertility would seem to make a strong case both for considering infertility as a disease and IVF as a legitimate treatment of that disease. Two contributors discuss the psychological consequences of infertility. Ponticas and Fagan contend that biological parenthood is a major developmental task of adulthood which if not achieved can result in a sense of inadequacy in gender and sex roles. They refer to the "pathos and emptiness of childlessness." Parenthood is an integral and expected role of successful adulthood. As a society, we place great value on having children as a way to maximize personal fulfillment and as a contribution to society. Rosenthal reflects a similar position. The inability to bear children when one wishes, she says, is considered a developmental crisis. The ability to reproduce is considered by psychiatrists as central to an individual's core gender identity, self-concept, and body image. When reproduction is not possible, what commonly results is anger, guilt, blame, sadness, depressive symptoms, at times loss of self-esteem and control over one's life. Interactions with others may be altered and sexual difficulties, including diminution of sexual desire, may occur. The goal of infertility treatment, including IVF, is to help restore a sense of body wholeness and completeness, to treat depression, enhance self-esteem, decrease guilt and blame, promote optimism and reality, and decrease feelings of loneliness and isolation.

These observations probably reflect the thinking not only of scientists and clinicians, but also of the general public. There are assumptions contained herein, however, which require examination and are challenged by Jung and Blake, in particular. Jung, for example, maintains that the desire to have children is both natural and socially constructed. It is the latter that is of some concern. An exaggerated view of the importance of parenting that is culturally nourished may create in couples an obsessive attitude toward having children, with negative consequences resulting from the inability to do so. It is not difficult to understand why infertility is unbearable when having children is frequently portrayed as the sole criterion for assessing the fruitfulness of marriage or a woman's personal and sexual success and fulfillment. While she acknowledges that infertility can have far reaching effects on life satisfaction, well-being, and psychological adjustment, Blake also wonders how much of this is the result of social context and construction of the psychological needs of infertile couples. Similarly, Tauer asks: does society impose an imperative of biological parenthood on married couples so that they feel coerced to resort to various reproductive technologies?

These observations raise a nest of questions that are far more profound and significant than the usual focus on the morality of the procedure of IVF and ET. For example, how much *is* the felt need to have children and one's own biological children natural and how much is socially constructed? Is the desire for parenthood always a positive thing? This raises further questions about reasons for having children and our view of children. Is it appropriate to have children to achieve one's own personal fulfillment or as a contribution to society? Does the focus on having children distract us from considering and pursuing other ways of achieving personal fulfillment, being fruitful and contributing to society? If parenthood is so essential to psycosocial development, what is to be said of couples who, for whatever reason, either by choice or necessity, remain childless? Are they to be considered failures, social misfits, irresponsible citizens? Unfortunately, for the most part, these questions are neglected in many if not most discussions of the moral justifiability of reproductive technologies. The Instruction does not address them nor do the contributors to this volume address them sufficiently. Yet they are critical to any moral assessment of IVF and other technological means of procreative assistance.

Related to the assumption about the significance of biological parenthood is another assumption about the role of technology in treating disabilities and disease. There is at times an attitude of technological pragmatism which pervades the scientific and health care communities. It is often referred to as the technological imperative: because something can be done, it should be done. There is a touch of some of this type of thinking in some of these pages. The reasoning goes something like this. Infertility inflicts serious psychosocial consequences on couples. We have technological means to compensate for this infertility and thus eliminate its threats to personal identity, well-being, and fulfillment and to the fruitfulness and success of the marriage. Hence, these means should be employed. If they can somehow fix the problem, they should be used to fix the problem. This is pure pragmatism. It is the classic means-ends issue. And clearly, not all means are morally justifiable to achieve the ends we want to achieve. In reality, it is not only the consequences that count, but also the way in which the consequences are achieved. So the claim that we have the means to effectively "treat" the effects of infertility is not sufficient. It does not settle the question. One must also address the moral justifiability of the methods employed. To the Instruction's credit, it does raise the issue of our attitudes toward medical technology, though one might question the tone of the discussion as well as its adequacy. One further point here. Infertility is undoubtedly a biological problem, but it is also psychological and cultural. IVF may address the biological and psychological dimensions, but it does not address the cultural and may in fact simply perpetuate conditions that contribute to the negative responses to infertility. IVF and other reproductive technologies may not be the appropriate "fix" for the deeper issues associated with infertility.

Another assumption related to the significance of biological parenthood and to the psychological consequences of infertility is that procreative decisions are solely individual decisions. This problem was already briefly discussed under methodology. Like so many, if not most issues in health care ethics, the moral justification of IVF is generally cast in individualistic rights terms or as a matter of individual health needs. Couples have a right to achieve conception in whatever manner they personally find feasible. For Blake, the most important moral issue regarding reproductive technologies is not the structure of the act, but the "unreflective detachment of individual/partner decision-making from the social, political, and economic context." It is the assumption that autonomous liberty, interpreted as a right of noninterference and freedom from social obligation, frees the pursuit of individual fulfillment from any relationship to and responsibility for the common good. This, according to Blake, is problematic in itself. But it is even more problematic for those within the Catholic tradition which is committed to a different assumption: respect for the inalienable dignity of the individual, but always within a social context and in relationship to the common good. Unfortunately, as was noted previously, *Donum Vitae* misses the opportunity to challenge to prevailing framework for discussions of reproductive technologies.

This emphasis on individual rights and needs, Blake, Jung and others believe, contribute to problems of social justice, also omitted from discusion in *Donum Vitae*. One such problem is access to reproductive technologies. Blake maintains that because infertility and its treatment are defined in personal terms, those who benefit from the availability of IVF tend to be white, upper-and-middle class, well-educated professionals. Those at the bottom of the social order have little or no access, yet have considerable need. There is little doubt, according to Blake, that socioeconomic status is a significant determinant of whose psychological, reproductive needs are met. And this inequity is perpetuated by marketplace strategies, which construct and normalize the psychological needs of infertile couples, and government policies. Blake even goes to the point of saying that by participating in the institution of reproductive technology, individuals are complicitous in the maintenance of a system that " focuses on individual self-interest to the neglect of the common good, releases government and social institutions from social responsibility for the common good . . . , and increases the disparity in reproductive privilege between the socioeconomically advantaged and disadvantaged."

Jung also discusses access to reproductive technologies, but she does so in relation to the issue of a prudent and just allocation of health care resources. She asks how society should weigh the claim of the 1 million people who sought treatment for infertility in 1990 against other competing demands for health care. She maintains that, in the face of the current inequitable distribution of health care resources where so many are excluded from basic care, good stewardship and the demands of justice require that tough decisions be made about health care priorities. One such decision, she believes, should be that public funding and institutional support for IVF be given low priority in comparison to other pressing unmet health care needs. This is additionally the case since IVF is expensive and often futile and the ratio between its benefits and burdens is poor. Jung believes that such a decision should not be construed as good, and that it can only be justified if the resources saved are used to improve everyone's access to more basic and beneficial services. Justice, Jung observes, does not come cheaply. At times, it comes at the cost of some individual liberties, and the burdens of implementing justice may be experienced more by some groups than others.

Finally, Tauer and several other authors raise the question whether societal expectations about parenthood and the promotion of IVF programs end up making women vulnerable to exploitation. Not only are women susceptible to manipulation in the use of IVF, but they are the ones who must carry the bulk of the burden even when the infertility problem is with the husband.

Discussion of these assumptions and their associated considerations is an unfortunate omission in *Donum Vitae's* treatment of IVF, particularly in view of the tradition's enormous concern for parenthood, human dignity, and social justice. But again, preoccupation with maintaining the integrity of the sexual act seems to have blinded the document's authors to equally if not more serious issues.

Perhaps this is an unfair charge to make for, as Porter points out, the Instruction "does not attempt a comprehensive presentation, or much less, a theoretical development of the magisterium's teachings on sexuality and reproduction. It simply applies those teachings . . . to questions that have been raised concerning new medical possibilities." The same may even be true for some of the charges made throughout these pages. And, in all fairness, if *Donum Vitae* can be criticized for not doing what it did not intend to do, then the same should be true of the contributions to this volume. One of the recurrent themes, by absence, is discussion of what is positive in the Instruction, where the Instruction's methodology, or argumentation, or assumptions make some contribution to the debate about IVF. In fact, only May makes a case in support of the Instruction's prohibition of homologous *in vitro* fertilization and embryo transfer based particularly on its argument of the "language of the body" and the "dignity of the child." Another omission is the lack of developed constructive responses to the Instruction's deficiencies. There are sometimes intimations of how one might or should think differently about a particular matter, but no sustained development of an alternative point of view. But then again, this may not have been the intent of the essay in question.

So where does all of this leave us? In the end, has the Instruction "had a great fall" and is it irreparably shattered? Can all the king's horses and all the king's men put Humpty together again? That remains to be seen. At minimum, *Donum Vitae* has not adequately and convincingly made its case. This would seem to be an essential requirement for an approach to moral teaching that claims to be based on reason. The only other way to gain compliance is by appeal to authority, and this will not be successful. In all probability, there are serious cracks in the Instruction's case against homologous IVF. The severity of those cracks must be left to the judgment of the reader. Of the various critiques levelled at the document, not all are of equal weight nor are all necessarily valid. Assessment of these critiques would require another volume or at least another essay. Furthermore, we have only one side of the debate. We do not have responses to the criticisms made. All the king's horses and all the king's men have not attempted to put Humptey together again.

Whether or not they do, there are critical lessons to be learned from *Donum Vitae* and from the numerous responses to it, including those in this volume. At least one such lesson is this: even the "simple case" is multi-faceted and highly complex. In order to be persuasive, moral analysis and argument must do justice both to the complexity of an issue and to its many dimensions. This will require not only careful thought, but also broad and sustained inquiry, dialogue, and debate. Short-circuiting these will only undermine the moral authority of the Church and the persuasive force of its teaching. It will have no impact on those matters of moral and social concern it seeks to address.

Another lesson I am left with is the amount of creative, constructive work which the Christian community and its theologians still have to do in the area of reproductive technology. And this work is not limited to a moral assessment of these various technologies. It must address underlying assumptions about such fundamental matters as the importance of parenthood, our views on infertility, what it means to have one's own biological child, the reasons for having children, how reproductive technologies fit into our health care system and what impact they have on women. Lacking a serious and thorough rethinking of these and other issues, our moral analyses of reproductive technologies will fall far short of what they need to be. In the final analysis, it might be best if all the king's horses and all the king's men can't put Humpty together again.

APPENDIX: SACRED CONGREGATION FOR THE DOCTRINE

INSTRUCTION ON RESPECT FOR HUMAN LIFE IN ITS ORIGIN AND ON THE DIGNITY OF PROCREATION

Replies to Certain Questions of the Day

FOREWORD

The Congregation for the Doctrine of the Faith has been approached by various Episcopal Conferences or individual Bishops, by theologians, doctors and scientists, concerning biomedical techniques which make it possible to intervene in the initial phase of the life of a human being and in the very processes of procreation and their conformity with the principles of Catholic morality. The present Instruction, which is the result of wide consultation and in particular of a careful evaluation of the declarations made by Episcopates, does not intend to repeat all the Church's teaching on the dignity of human life as it originates and on procreation, but to offer, in the light of the previous teachings of the Magisterium, some specific replies to the main questions being asked in this regard.

The expositions will be arranged as follows: an introduction will recall the fundamental principles, of an anthropological and moral character, which are necessary for a proper evaluation of the problems and for working out replies to those questions; the first part will have as its subject respect for the human being from the first moment of his or her existence; the second part will deal with the moral questions raised by technical interventions on human procreation; the third part will offer some orientations on the relationships between moral law and civil law in terms of the respect due to human embryos and foetuses and as regards the legitimacy of techniques of artificial procreation.

INTRODUCTION

1. Biomedical Research and the Teaching of the Church

The gift of life which God the Creator and Father has entrusted to man calls him to appreciate the inestimable value of what he has been given and to take responsibility for it: this fundamental principle must be placed at the centre of one's reflection in order to clarify and solve the moral problems raised by artificial interventions on life as it originates and on the processes procreation.

Thanks to the progress of the biological and medical sciences, man has at his disposal ever more effective therapeutic resources; but he can also acquire new powers, with unforeseeable consequences, over human life at its very beginning and in its first stages. Various procedures now make it possible to intervene not only in order to assist but also to dominate the processes of procreation. These techniques can enable man to "take in his hand his own destiny", but they also expose him "to the temptation to go beyond the limits of a reasonable dominion over nature".[1] They might constitute progress in the service of man, but they also involve serious risks. Many people are therefore expressing an urgent appeal that in interventions on procreation the values and rights of the human person be safeguarded. Requests for clarification and guidance are coming not only from the faithful but also from those who recognize the Church as "an expert in humanity"[2] with a mission to serve the "civilization of love"[3] and of life.

The Church's Magisterium does not intervene on the basis of a particular competence in the area of the experimental sciences; but having taken account of the data of research and technology, it intends to put forward, by virtue of its evangelical mission and apostolic duty, the moral teaching corresponding to the dignity of the person and to his or her integral vocation. It intends to do so by expounding the criteria of moral judgement as regards the applications of scientific research and technology, especially in relation to human life and its beginnings. These criteria are the respect, defence and promotion of man, his "primary and fundamental right" to life,[4] his dignity as a person who is endowed with a spiritual soul and with moral responsibility[5] and who is called to beatific communion with God.

The Church's intervention in this field is inspired also by the love which she owes to man, helping him to recognize and respect his rights and duties. This love draws from the fount of Christ's love: as she contemplates the mystery of the Incarnate Word, the Church also comes to understand the "mystery of man";[6] by proclaiming the Gospel of salvation, she reveals to man his dignity and invites him to discover fully the truth of his own being. Thus the Church once more puts forward the divine law in order to accomplish the work of truth and liberation.

For it is out of goodness – in order to indicate the path of life – that God gives human beings his commandments and the grace to observe them: and it is likewise out of goodness – in order to help them persevere along the same path – that God always offers to everyone his forgiveness. Christ has compassion on our weaknesses: he is our Creator and Redeemer. May his spirit open men's hearts to the gift of God's peace and to an understanding of his precepts.

2. Science and Technology at the Service of the Human Person

God created man in his own image and likeness: "male and female he created them (*Gen* 1:27), entrusting to them the task of "having dominion over the earth" (*Gen* 1:28). Basic scientific research and applied research constitute a significant expression of this dominion of man over creation. Science and technology are valuable resources for man when placed at his service and when they promote his integral development for the benefit of all; but they cannot of themselves show the meaning of existence and of human progress. Being ordered to man, who initiates and develops them, they draw from the person and his moral values the indication of their purpose and the awareness of their limits.

It would on the one hand be illusory to claim that scientific research and its applications are morally neutral; on the other hand one cannot derive criteria for guidance from mere technical efficiency, from research's possible usefulness to some at the expense of others, or, worse still, from prevailing ideologies. Thus science and technology require, for their own intrinsic meaning, an unconditional respect for the fundamental criteria of the moral law: that is to say, they must be at the service of the human person, of his inalienable rights and his true and integral good according to the design and will of God.[7]

The rapid development of technological discoveries gives greater urgency to this need to respect the criteria just mentioned: science without conscience can only lead to man's ruin. "Our era needs such wisdom more than bygone ages if the discoveries made by man are to be further humanized. For the future of the world stands in peril unless wiser people are forthcoming".[8]

3. Anthropology and Procedures in the Biomedical Field

Which moral criteria must be applied in order to clarify the problems posed today in the field of biomedicine? The answer to this question presupposes a proper idea of the nature of the human person in his bodily dimension.

For it is only in keeping with his true nature that the human person can achieve self-realization as a "unified totality":[9] and this nature is at the same time corporal and spiritual. By virtue of its substantial union with a spiritual soul, the human body cannot be considered as a mere complex of tissues, organs and functions, nor can it be evaluated in the same way as the body of animals; rather it is a constitutive part of the person who manifests and expresses himself through it.

The natural moral law expresses and lays down the purposes, rights and duties which are based upon the bodily and spiritual nature of the human person. Therefore this law cannot be thought of as simply a set of norms on the biological level; rather it must be defined as the rational order whereby man is called by the Creator to direct and regulate his life and actions and in particular to make use of his own body.[10]

A first consequence can be deduced from these principles : an intervention on the human body not only affects the tissues, the organs and their functions but also involves the person himself on different levels. It involves, therefore, perhaps in an implicit but nonetheless real way, a moral significance and responsibility. Pope John Paul II forcefully reaffirmed this to the World Medical Association when he said: "Each human person, in his absolutely unique singularity, is constituted not only by his spirit, but by his body as well. Thus, in the body and through the body, one touches the person himself in his concrete reality. To respect the dignity of man consequently amounts to safeguarding this identity of the man *'corpore et anima unus'*, as the Second Vatican Council says (*Gaudium et Spes*, 14, par. 1). It is on the basis of this anthropological vision that one is to find the fundamental criteria for decision-making in the case of procedures which are not to strictly therapeutic, as, for example, those aimed at the improvement of the human biological condition".[11]

Applied biology and medicine work together for the integral good of human life when they come to the aid of a person stricken by illness and infirmity and when they respect his or her dignity as a creature of God. No biologist or doctor can reasonably claim, by virtue of his scientific competence, to be able to decide on people's origin and destiny. This norm must be applied in a particular way in the field of sexuality and procreation, in which man and woman actualize the fundamental values of love and life.

God, who is love and life, has inscribed in man and woman the vocation to share in a special way in his mystery of personal communion and in his work as Creator and Father.[12] For this reason marriage possesses specific goods and values in its union and in procreation which cannot be likened to those existing in lower forms of life. Such values and meanings are of the personal order and determine from the moral point of view the meaning and limits of artificial interventions on procreation and on the origin of human life. These interventions are not to be rejected on the grounds that they are artificial. As such, they bear witness to the possibilities of the art of medicine. But they must be given a moral evaluation in reference to the dignity of the human person, who is called to realize his vocation from God to the gift of love and the gift of life.

4. Fundamental Criteria for a Moral Judgement

The fundamental values connected with the techniques of artificial human procreation are two: the life of the human being called into existence and the special nature of the transmission of human life in marriage. The moral judgement on such methods of artificial procreation must therefore be formulated in reference to these values.

Physical life, with which the course of human life in the world begins, certainly does not itself contain the whole of a person's value, nor does it represent the supreme good of man who is called to eternal life. However it does constitute in a certain way the "fundamental" value of life, precisely because upon this physical life all the other values of the person are based and developed.[13] The inviolability of the innocent human being's right to life "from the moment of conception until death"[14] is a sign and requirement of the very inviolability of the person to whom the Creator has given the gift of life.

By comparison with the transmission of other forms of life in the universe, the transmission of human life has a special character of its own, which derives from the special nature of the human person. "The transmission of human life is entrusted by nature to a personal and conscious act and as such is subject to the all-holy laws of God: immutable and inviolable laws which must be recognized and observed. For this reason one cannot use means and follow methods which could be licit in the transmission of the life of plants and animals".[15]

Advances in technology have now made it possible to procreate apart from sexual relations through the meeting *in vitro* of the germ-cells previously taken from the man and the woman. But what is technically possible is not for that very reason morally admissible. Rational reflection on the fundamental values of life and of human procreation is therefore indispensable for formulating a moral evaluation of such technological interventions on a human being from the first stages of his development.

5. *Teachings of the Magisterium*

On its part, the Magisterium of the Church offers to human reason in this field too the light of Revelation: the doctrine concerning man taught by the Magisterium contains many elements which throw lights on the problems being faced here.

From the moment of conception, the life of every human being is to be respected in an absolute way because man is the only creature on earth that God has "wished for himself"[16] and the spiritual soul of each man is 'immediately created" by God;[17] his whole being bears the image of the Creator. Human life is sacred because from its beginning it involves "the creative action of God"[18] and it remains forever in a special relationship with the Creator, who is its sole end.[19] God alone is the Lord of life from its beginning until its end: no one can, in any circumstance, claim for himself the right to destroy directly an innocent human being.[20]

Human procreation requires on the part of the spouses responsible collaboration with the fruitful love of God;[21] the gift of human life must be actualized in marriage through the specific and exclusive acts of husband and wife, in accordance with the laws inscribed in their persons and in their union.[22]

I. RESPECT FOR HUMAN EMBRYOS

Careful reflection on this teaching of the Magisterium and on the evidence of reason, as mentioned above, enables us to respond to the numerous moral problems posed by technical interventions upon the human being in the first phases of his life and upon the processes of his conception.

1. What Respect Is Due to the Human Embryo, Taking into Account His Nature and Identity?

> The human being must be respected – as a person – from the very first instant of his existence.

The implementation of procedures of artificial fertilization has made possible various interventions upon embryos and human foetuses. The aims pursued are of various kinds: diagnostic and therapeutic, scientific and commercial. From all of this, serious problems arise. Can one speak of a right to experimentation upon human embryos for the purpose of scientific research? What norms or laws should be worked out with regard to this manner? The response to these problems presupposes a detailed reflection on the nature and specific identity – the word "status" is used – of the human embryo itself.

At the Second Vatican Council, the Church for her part presented once again to modern man her consent and certain doctrine according to which: "Life once conceived, must be protected with the utmost care; abortion and infanticide are abominable crimes".[23] More recently, the *Charter of the Rights of the Family*, published by the Holy See, confirmed that "Human life must be absolutely respected and protected from the moment of conception".[24]

This Congregation is aware of the current debates concerning the beginning of human life, concerning the individuality of the human being and concerning the identity of the human person. The Congregation recalls the teachings found in the *Declaration on Procured Abortion*: "From the time that the ovum is fertilized, a new life is begun which is neither that of the father nor of the mother; it is rather the life of a new human being with his own growth. It would never be made human if it were not human already. To this perpetual evidence . . . modern genetic science brings valuable confirmation. It has demonstrated that, form the first instant, the programme is fixed as to what this living being will be: a man, this individual-man with his characteristic aspects already well determined. Right from fertilization is begun the adventure of a human life, and each of its great capacities requires time . . . to find its place and be in a position to act".[25] This teaching remains valid and is further confirmed, if confirmation were needed, by recent findings of human biological science which recognize that in the zygote resulting from fertilization the biological identity of a new human individual is already constituted.

Certainly no experimental datum can be in itself sufficient to bring us to the recognition of a spiritual soul; nevertheless, the conclusions of science regarding the human embryo provide a valuable indication for discerning by the use of reason a personal presence at the moment of this first appearance of a human life: how could a human individual not be a human person? The Magisterium has not expressly committed itself to an affirmation of a philosophical nature, but it constantly reaffirms the moral condemnation of any kind of procured abortion. This teaching has not been changed and is unchangeable.[26]

Thus the fruit of human generation, from the first moment of its existence, that is to say from the moment the zygote has formed, demands the unconditional respect that is morally due to the human being in his bodily and spiritual totality. The human being is to be respected and treated as a person from the moment of conception; and therefore from that same moment his rights as a person must be recognized, among which in the first place is the inviolable right of every innocent human being to life.

This doctrinal reminder provides the fundamental criterion for the solution of the various problems posed by the development of the biomedical science in this field: since the embryo must be treated as a person, it must also be defended in its integrity, tended and cared for, to the extent possible, in the same way as any other human being as far as medical assistance is concerned.

2. Is Prenatal Diagnosis Morally Licit?

> If prenatal diagnosis respects the life and integrity of the embryo and the human foetus and is directed towards its safeguarding or healing as an individual, then the answer is affirmative.

For prenatal diagnosis makes it possible to know the condition of the embryo and foetus when still in the mother's womb. It permits, or makes it possible to anticipate earlier and more effective, certain therapeutic, medical or surgical procedures.

Such diagnosis is permissible, with the consent of the parents after they have been adequately informed, if the methods employed safeguard the life and integrity of the embryo and the mother, without subjecting them to disproportionate risks.[27] But this diagnosis is gravely opposed to the moral law when it is done with the thought of possibly inducing an abortion depending on the results: a diagnosis which shows the existence of a malformation or a hereditary illness must not be the equivalent of a death-sentence. Thus a woman would be committing a gravely illicit act if she were to request such a diagnosis with the deliberate intention of a having an abortion should the results confirm the existence of a malformation or abnormality. The spouse or relatives or anyone else would similarly be acting in a manner contrary to the moral law of they were to counsel the expectant mother with the same intention of possibly proceeding to an abortion. So too the specialist would be guilty of illicit collaboration if, in conducting the diagnosis and in communicating its results, he were deliberately to contribute to establishing or favouring a link between prenatal diagnosis and abortion.

In conclusion, any directive or programme of the civil and health authorities or of scientific organizations which in any way were to favour a link between prenatal diagnosis and abortion, or which were to go as far as directly to induce expectant mothers to submit to prenatal diagnosis planned for the purpose of eliminating foetuses which are affected by malformations or which are carriers of hereditary illness, is to be condemned as a violation of the unborn child's rights to life and as an abuse of the prior rights and duties of the spouses.

3. Are Therapeutic Procedures Carried Out on the Human Embryo Licit?

> *As with all medical interventions on patients*, one must uphold as licit procedures carried out on the human embryo which respect the life and integrity of the embryo and do not involve disproportionate risks for it but are directed towards its healing, the improvement of its condition of health, or its individual survival.

Whatever the type of medical, surgical or other therapy, the free and informed consent of the parents is required, according to the deontological rules followed in the case of children. The application of this moral principle may call for delicate and particular precautions in the case of embryonic or foetal life.

The legitimacy and criteria of such procedures have been clearly stated by Pope John Paul II: "A strictly therapeutic intervention whose explicit objective is the healing of various maladies such as those stemming from chromosomal defects will, in principle, be considered desirable, provided it is directed to the true promotion of the personal well-being of the individual without doing harm to his integrity or worsening his conditions of life. Such an intervention would indeed fall within the logic of the Christian moral tradition".[28]

4. How Is One to Evaluate Morally Research and Experimentation on Human Embryos and Foetuses?

> Medical research must refrain form operations on live embryos, unless there is a moral certainty of not causing harm to the life or integrity of the unborn child and the mother, on the condition that the parents have given their free and uninformed consent to the procedure.

It follows that all research, even when limited to the simple observation of the embryo, would become illicit were it to involve risk to the embryo's physical integrity or life by reason of the methods used or the effects induced.

As regards experimentation, and presupposing the general distinction between experimentation for purposes which are not directly therapeutic and experimentation which is clearly therapeutic for the subject himself, in the case in point one must also distinguish between experimentation carried out on embryos which are dead. *If the embryos are living, whether viable or not, they must be respected just like any other human person; experimentation on embryos which is not directly therapeutic is illicit.*[29]

No objective, even though noble in itself, such as a foreseeable advantage to science, to other human beings or to society, can in any way justify experimentation on living human embryos or foetuses, whether viable or not, either inside or outside the mother's womb. The informed consent ordinarily required for clinical experimentation on adults cannot be granted by the parents, who may not freely dispose of the physical integrity or life of the unborn child. Moreover, experimentation on embryos and foetuses always involves risk, and indeed in most cases it involves the certain expectation of harm to their physical integrity or even their death.

To use human embryos or foetuses as the object or instrument of experimentation constitutes a crime against their dignity as human beings having a right to the same respect that is due to the child already born and to every human person.

The *Charter of the Rights of the Family* published by the Holy See affirms: "Respect for the dignity of the human being excludes all experimental manipulation or exploitation of the human embryo".[30] The practice of keeping alive human embryos *in vivo* or *in vitro* for experimental or commercial purposes it totally opposed to human dignity.

In the case of experimentation that is clearly therapeutic, namely, when it is a matter of experimental forms of therapy used for the benefit of the embryo itself in a final attempt to save its life, and in the absence of other reliable forms of therapy, recourse to drugs or procedures not yet fully tested can be licit.[31]

The corpses of human embryos and foetuses, whether they have been deliberately aborted or not, must be respected just as the remains of other human beings. In particular, they cannot be subjected to mutilation or to autopsies without the consent of the parents or of the mother. Furthermore, the moral requirements must be safeguarded that there be no complicity in deliberate abortion and that the risk of scandal be avoided. Also, in the case of dead foetuses, as for the corpses of adult persons, all commercial trafficking must be considered illicit and should be prohibited.

5. *How Is One to Evaluate Morally the Use for Research Purposes of Embryos Obtained by Fertilization 'In Vitro'?*

Human embryos obtained *in vitro* are human beings and subjects with rights: their dignity and right to life must be respected from the first moment of their existence. *It is immoral to produce human embryos destined to be exploited as disposable "biological material".*

In the usual practice of *in vitro* fertilization, not all of the embryos are transferred to the woman's body; some are destroyed. Just as the Church condemns induced abortion, so she also forbids acts against the life of these human beings. *It is a duty to condemn the particular gravity of the voluntary destruction of human embryos obtained 'in vitro' for the sole purpose of research, either by means of artificial insemination or by means of "twin fission".* By acting in this way the researcher usurps the place of God; and, even though he may be unaware of this, he sets himself up as the master of the destiny of others inasmuch as he arbitrarily chooses whom he will allow to live and whom he will send to death and kills denfenceless human beings.

Methods of observation or experimentation which damage or impose grave and disproportionate risks upon embryos obtained *in vitro* are morally illicit for the same reasons. Every human being is to be respected for himself, and cannot be reduced in worth to a pure and simple instrument for the advantage of others. *It is therefore not in conformity with the moral law deliberately to expose to death human embryos obtained 'in vitro'.* In consequence of the fact that they have been produced *in vitro*, those embryos which art not transferred into the body of the mother and are called "spare" are exposed to an absurd fate, with no possibility of their being offered safe means of survival which can be licitly pursued.

6. What Judgement should be Made on Other Procedures of Manipulating Embryos Connected with the "Techinques of Human Reproduction"?

Techniques of fertilization *in vitro* can open the way to other forms of biological and genetic manipulation of human embryos, such as attempts or plans for fertilization between human and animal gametes and the gestation of human embryos in the uterus of animals, or the hypothesis or project of constructing artificial uteruses for the human embryo. *These procedures are contrary to the human dignity proper to the embryo, and at the same time they are contrary to the right of every person to be conceived and to be born within marriage and from marriage.*[32] *Also, attempts or hypotheses for obtaining a human being without any connection with sexuality through "twin fission", cloning or parthenogenesis are to be considered contrary to moral law, since they are in opposition to the dignity of both human procreation and of the conjugal union.*

The freezing of embryos, even when carried out in order to preserve the life of an embryo – cryopreservation – *constitutes an offence against the respect due to human beings* by exposing them to grave risks of death or harm to their physical integrity and depriving them, at least temporarily, of maternal shelter and gestation, thus placing them in a situation in which further offences and manipulation are possible.

Certain attempts to influence chromosomic or genetic inheritance are not therapeutic but are aimed at producing human beings selected according to sex or other predetermined qualities. These manipulations are contrary to the personal dignity of the human being and his or her integrity and identity. Therefore in no way can they be justified on the grounds of possible beneficial consequences for future humanity.[33] Every person must be respected for himself: in this consists the dignity and right of every human being from his or her beginning.

II. INTERVENTIONS UPON HUMAN PROCREATION

By "artificial procreation" or "artificial fertilization" are understood here the different technical procedures directed towards obtaining a human conception in a manner other than the sexual union of man and woman. This Instruction deals with fertilization of an ovum in a test-tube (*in vitro* fertilization) and artificial insemination through transfer into the woman's genital tracts of previously collected sperm.

A preliminary point for the moral evaluation of such technical procedures is constituted by the consideration of the circumstances and consequences which those procedures involve in relation to the respect due the human embryo. Development of the practice of *in vitro* fertilization has required innumerable fertilizations and destructions of human embryos. Even today, the usual practice presuppose a hyper ovulation on the part of the woman: a number of ova are withdrawn, fertilized and then cultivated *in vitro* for some days. Usually not all are transferred into the genital tracts of the woman; some embryos, generally called "spare", are destroyed or frozen. On occasion, some of the implanted embryos are sacrificed for various eugenic, economic or psychological reasons. Such deliberate destruction of human beings or their utilization for different purposes to the detriment of their integrity and life is contrary to the doctrine on procured abortion already recalled.

The connection between *in vitro* fertilization and the voluntary destruction of human embryos occurs too often. This is significant: through these procedures, with apparently contrary purposes, life and death are subjected to the decision of man, who thus sets himself up as the giver of life and death by decree. This dynamic of violence and domination may remain unnoticed by those very individuals who, in wishing to utilize this procedure, become subject to it themselves. The facts recorded and the cold logic which links them must be taken into consideration for a moral judgement on IVF and ET (*in vitro* fertilization and embryo transfer): the abortion-mentality which has made this procedure possible thus leads, whether one wants it or not, to man's domination over the life and death of his fellow human beings and can lead to a system of radical eugenics.

Nevertheless, such abuses do not exempt one from a further and thorough ethical study of the techniques of artificial procreation considered in themselves, abstracting as far as possible from the destruction of embryos produced *in vitro*.

The present Instruction will therefore take into consideration in the first place the problems posed by heterologous artificial fertilization (II, 1–3), and subsequently those linked with homologous artificial fertilization (II, 4–6).

Before formulating an ethical judgement on each of these procedures, the principles and values which determine the moral evaluation of each of them will be considered.

A. Heterologous Artificial Fertilization

1. Why Must Human Procreation Take Place in Marriage?

> Every human being is always accepted as a gift and blessing of God. However, from the moral point of view a truly responsible procreation vis-a-vis the unborn child must be the fruit of marriage.

For human procreation has specific characteristics by virtue of the personal dignity of the parents and of the children: the procreation of a new person, whereby the man and the woman collaborate with the power of the Creator, must be the fruit and the sign of the mutual self-giving of the spouses, of their love and of their fidelity.[34] *The fidelity of the spouses in the unity of marriage involves reciprocal respect of their right to become a father and a mother only through each other.*

The child has the right to be conceived, carried in the womb, brought into the world and brought up within marriage: it is through the secure and recognized relationship to his own parents that the child can discover his own identity and achieve his own proper human development.

The parents find in their child a confirmation and completion of their reciprocal self-giving: the child is the living image of their love, the permanent sign of their union, the living and indissoluble concrete expression of their paternity and maternity.[35]

By reason of the vocation and social responsibilities of the person, the good of the children and of the parents contributes to the good of civil society; the vitality and stability of society require that the family be firmly based on marriage.

The tradition of the Church and anthropological reflection recognize in marriage and in its indissoluble unity the only setting worthy of truly responsible procreation.

2. Does Heterologous Artificial Fertilization Conform to the Dignity of the Couple and to the Truth of Marriage?

Through IVF and ET and heterologous artificial insemination, human conception is achieved through the fusion of gametes of at least one donor other than the spouses united in marriage. *Heterologous artificial fertilization is contrary to the unity of marriage, to the dignity of the spouses, to the vocation proper to parents, and to the child's right to be conceived and brought into the world in marriage and from marriage.*[36]

Respect for the unity of marriage and for conjugal fidelity demands that the child be conceived in marriage; the bond existing between husband and wife accords the spouses, in an objective and inalienable manner, the exclusive right to become father and mother solely through each other.[37] Recourse to the gametes of a third person, in order to have sperm or ovum available, constitutes a violation of the reciprocal commitment of the spouses and a grave lack in regard to that essential property of marriage which is its unity.

Heterologous artificial fertilization violates the rights of the child; it deprives him of his filial relationship with his parental origins and can hinder the maturing of his personal identity. Furthermore, it offends the common vocation of the spouses called to fatherhood and motherhood: it objectively deprives conjugal fruitfulness of its unity and integrity; it brings about and manifests a rupture between genetic parenthood, gestational parenthood and responsibility for upbringing. Such damage to the personal relationships within the family has repercussions on civil society: what threatens the unity and stability of the family is a source of dissension, disorder and injustice in the whole of social life.

These reasons lead to a negative moral judgment concerning heterologous artificial fertilization: consequently fertilization of a married woman with the sperm of a donor different from her husband and fertilization with the husband's sperm of an ovum not coming from his wife are morally illicit. Furthermore, the artificial fertilization of a woman who is unmarried or a widow, whoever the donor may be, cannot be morally justified.

The desire to have a child and the love between spouses who long to obviate a sterility which cannot be overcome in any other way constitute understandable motivations; but subjectively good intentions do not render heterologous artificial fertilization conformable to the objective and inalienable properties of marriage or respectful pf the rights of the child and of the spouses.

3. Is "Surrogate" Motherhood Morally Licit?

> No, for the same reasons which lead one to reject heterologous artificial fertilization: for it is contrary to the unity of marriage and to the dignity of the procreation of the human person.

Surrogate motherhood represents an objective failure to meet the obligations of maternal love, of conjugal fidelity and of responsible motherhood; it offends the dignity and the right of the child to be conceived, carried in the womb. Brought into the world and brought up by his own parents; it sets up, to the detriment of families, a division between the physical, psychological and moral elements which constitute those families.

B. Homologous Artificial Fertilization

Since heterologous artificial fertilization has been declared unacceptable, the question arises of how to evaluate morally the process of homologous artificial fertilization: IVF and ET and artificial insemination between husband and wife. First a question of principle must be clarified.

4. What Connection Is Required from the Moral Point of View Between Procreation and the Conjugal Act?

A. The Church's teaching on marriage and human procreation affirms the "inseparable connection. Willed by God and unable to be broken by man on his own initiative, between the two meanings of the conjugal act: the unitive meaning and the procreative meaning. Indeed, by its intimate structure, the conjugal act, while most closely uniting husband and wife, capacitates them for the generation of new lives, according to laws inscribed in the very being of man and of woman".[38] This principle, which is based upon the nature of marriage and the intimate connection of the goods of marriage, has well-known consequences on the level of responsible fatherhood and motherhood. "By safeguarding both these essential aspects, the unitive and the procreative, the conjugal act preserves in its fullness the sense of true mutual love and its ordination towards man's exalted vocation to parenthood".[39]

The same doctrine concerning the link between the meanings of the conjugal act and between the goods of marriage throws light on the moral problem of homologous artificial fertilization, since "it is never permitted to separate these different aspects to such a degree as positively to exclude either the procreative intention or the conjugal relation".[40]

Contraception deliberately deprives the conjugal act of its openness to procreation and in this way brings about a voluntary dissociation of the ends of marriage. Homologous artificial fertilization, in seeking a procreation which is not the fruit of a specific act of conjugal union, objectively effects an analogous separation between the goods and the meanings of marriage.

Thus, *fertilization is licitly sought when it is the result of a "conjugal act which is per se suitable for the generation of children to which marriage is ordered by its nature and by which the spouses become one flesh".*[41] *But from the moral point of view procreation is deprived of its proper perfection when it is not desired for the fruit of the conjugal act, that is to say of the specific act of the spouses' union.*

B. The moral value of the intimate link between the goods of marriage and between the meanings of the conjugal is based upon the unity of a human being, a unity involving body and spiritual soul.[42] Spouses mutually express their personal love in the "language of the body", which clearly involves both "sponsal meanings" and parental ones.[43] The conjugal act by which the couple mutually express their self-gift at the same time expresses one openness to the gift of life. It is an act that is inseparably corporal and spiritual. In order to respect the language of their bodies and their mutual generosity, the conjugal union must take place with respect for its openness to procreation; and the procreation of a person must be the fruit and the result of married love. The origin of the human being thus follows form a procreation that is "linked to the union, not only biological but also spiritual, of the parents, made one by the bond of marriage".[44] Fertilization achieved outside the bodies of the couples remains by this very fact deprived of the meanings and the values which are expressed in the language of the body and in the union of human persons.

C. Only respect for the link between the meanings of the conjugal act and respect for the unity of the human being make possible procreation in conformity with the dignity of the person. In his unique and irrepeatable origin, the child must be respected and recognized as equal in personal dignity to those who give him life. The human person must be accepted in his parents' act of union and love; the generation of a child must be therefore be the fruit of that mutual giving[45] which is realized in the conjugal act wherein the spouses cooperate as servants and not as masters in the work of the Creator who is Love.[46]

In reality, the origin of a human person is the result of an act of giving. The one conceived must be the fruit of his parents' love. He cannot be desired or conceived as the product of an intervention of medical or biological techniques; that would be equivalent to reducing him to an object of scientific technology. No one may subject the coming of a child into the world to conditions of technical efficiency which are to be evaluated according to standards of control and dominion.

The moral relevance of the link between the meanings of the conjugal act and between the goods of marriage, as well as the unity of the human being and the dignity of his origin, demand that the procreation of a human person be brought about as the fruit of the conjugal act specific to the love between spouses. The link between procreation and the conjugal act is thus shown to be of great importance on the anthropological and moral planes, and it throws light on the positions of the Magisterium with regard to homologous artificial fertilization.

5. Is Homologous 'In Vitro' Fertilization Morally Licit?

The answer to this question is strictly dependent on the principles just mentioned. Certainly one cannot ignore the legitimate aspirations of sterile couples. For some, recourse to homologous IVF and ET appears to be the only way of fulfilling their sincere desire for a child. The question is asked whether the totality of conjugal life in such situations is not sufficient to ensure the dignity proper to human procreation. It is acknowledged that IVF and ET certainly cannot supply for the absence of sexual relations[47] and cannot be preferred to the specific acts of conjugal union, given the risks involved for the child and the difficulties of the procedure. But it is asked whether, when there is no other way of overcoming the sterility which is a source of suffering, homologous *in vitro* fertilization may not constitute an aid, if not a form of therapy, whereby its moral licitness could be admitted.

The desire for a child – or at the very least an openness to the transmission of life – it is a necessary prerequisite from the moral point of view for responsible human procreation. But this good intention is not sufficient for making a positive moral evaluation of *in vitro* fertilization between spouses. The process of IVF and ET must be judged in itself and cannot borrow its definitive moral quality from the totality of conjugal life of which it becomes part nor from the conjugal acts which may precede or follow it.[48]

It has already been recalled that, in the circumstances in which it is regularly practised, IVF and ET involves the destruction of human beings, which is something contrary to the doctrine on the illicitness of abortion previously mentioned.[49] But even in a situation in which every precaution were taken to avoid the death of human embryos, homologous IVF and ET dissociates from the conjugal act the actions which are directed to human fertilization. For this reason the very nature of homologous IVF and ET also must be taken into account, even abstracting from the link with procured abortion.

Homologous IVF and ET is brought about outside the bodies of the couple through actions of third parties whose competence and technical activity determine the success of the procedure. Such fertilization entrusts the life and identity of the embryo into the power of the doctors and biologists and establishes the domination of technology over the origin and destiny of the human person. Such a relationship of domination is in itself contrary to the dignity and equality that must be common to parents and children.

Conception *in vitro* is the result of the technical action which presides over fertilization. *Such fertilization is neither in fact achieved nor positively willed as the expression and fruit of a specific act of the conjugal union. In homologous IVF and ET, therefore, even if it is considered in the context of 'de facto' existing sexual relations, the generation of the human person is objectively deprived of its proper perfection: namely, that of being the result and fruit of a conjugal act* in which the spouses can become "cooperators with God for giving life to a new person".[50]

These reasons enable us to understand why the act of conjugal love is considered in the teaching of the Church as the only setting worthy of human procreation. For the same reasons the so-called "simple case", i.e. a homologous IVF and ET procedure that is free of any compromise with the abortive practice of destroying embryos and with masturbation, remains a technique which is morally illicit because it deprives human procreation of the dignity which is proper and connatural to it.

Certainly, homologous IVF and ET fertilization is not marked by all the ethical negativity found in extra-conjugal procreation; the family and marriage continue to constitute the setting for the birth and upbringing of the children. Nevertheless, in conformity with the traditional doctrine relating to the goods of marriage and the dignity of the person, *the Church remain opposed from the moral point of view to homologous 'in vitro' fertilization. Such fertilization is in itself illicit and in opposition to the dignity of procreation and of the conjugal union, even when everything is done to avoid the death of the human embryo.*

Although the manner in which human conception is achieved with IVF and ET cannot be approved, every child which comes into the world must in any case be accepted as a living gift of the divine Goodness and must be brought up with love.

6. *How Is Homologous Artificial Insemination to be Evaluated from the Moral Point of View?*

> Homologous artificial insemination within marriage cannot be admitted except for those cases which the technical means is not a substitute foe the conjugal act but serves to facilitate and to help so that the act attains its natural purpose.

The teaching of the Magisterium on this point has already been stated.[51] This teaching is not just an expression of particular historical circumstances but is based on the Church's doctrine concerning the connection between the conjugal union and procreation and on a consideration of the personal nature of the conjugal act and of human procreation. "In its natural structure, the conjugal act is a personal action, a simultaneous and immediate cooperation on the part of the husband and wife, which by the very nature of the agents and the proper nature of the act is the expression of the mutual gift which, according to the words of Scripture, brings about union 'in one flesh' ".[52] Thus moral conscience "does not necessarily proscribe the use of certain artificial means destined solely either to the facilitating of the natural act or to ensuring that the natural act normally performed achieves its proper end".[53] If the technical means facilitates the conjugal act or helps it to reach its natural objectives, it can be morally acceptable. If, on the other hand, the procedure were to replace the conjugal act, it is morally illicit.

Artificial insemination as a substitute for the conjugal act is prohibited by reason of the voluntarily achieved dissociation of the two meanings of the conjugal act. Masturbation, through which the sperm is normally obtained, is another sign of this dissociation: even when it is done for the purpose of procreation, the act remains deprived of its unitive meaning: "It lacks the sexual relationship called for by the moral order, namely the relationship which realizes 'the full sense of mutual self-giving and human procreation in the context of true love' ".[54]

7. What Moral Criterion Can be Proposed with Regard to Medical Intervention in Human Procreation?

The medical act must be evaluated not only with reference to its technical dimension but also and above all in relation to its goal which is the good of persons and their bodily and psychological health. The moral criteria for medical intervention in procreation are deduced from the dignity of human persons, of their sexuality and of their origin.

Medicine which seeks to be ordered to the integral good of the person must respect the specifically human values of sexuality.[55] *The doctor is at the service of persons and of human procreation. He does not have the authority to dispose of them or to decide their fate.* A medical intervention respects the dignity of persons when it seeks to assist the conjugal act in order to facilitate its performance or in order to enable it to achieve its objective once it has been normally performed".[56]

On the other hand, to sometimes happens that a medical procedure technologically replaces the conjugal act in order to obtain a procreation which is neither its result nor its fruit. In this case the medical act is not, as it should be, at the service of conjugal union but rather appropriates to itself the procreative function and thus contradicts the dignity and the inalienable rights of the spouses and of the child to be born.

The humanization of medicine, which is insisted upon today by everyone, requires respect for the integral dignity of the human person first of all in the act and at the moment in which the spouses transmit life to a new person. It is only logical therefore to address an urgent appeal to Catholic doctors and scientists that they bear exemplary witness to the respect due to the human embryo and to the dignity of procreation. The medical and nursing staff of Catholic hospital and clinics are in a special way urged to do justice to the moral obligations which they have assumed, frequently also, as a part of their contract. Those who are in charge of Catholic hospitals and clinics and who are often Religious will take special care to safeguard and promote a diligent observance of the moral norms recalled in the present Instruction.

8. *The Suffering Caused by Infertility in Marriage*

> The suffering of spouses who cannot have children or who are afraid of bringing a handicapped child into the world is a suffering that everyone must understand and properly evaluate.

On the part of the spouses, the desire for a child is natural: it expresses the vocation to fatherhood and motherhood inscribed in conjugal love. This desire can be even stronger if the couple is affected by sterility which appears incurable. Nevertheless, marriage does not confer upon the spouses the right to have a child, but only the right to perform those natural acts which are *per se* ordered to procreation.[57]

A true and proper right to a child would be contrary to the child's dignity and nature. The child is not an object to which one has a right, nor can it be considered as an object of ownership: rather, a child is a gift, "the supreme gift" [58] *and the most gratuitous gift of marriage, and is living testimony of the mutual giving of his parents. For this reason, the child has the right, as already mentioned, to be the fruit of the specific act of the conjugal love of his parents; and he also has the right to be respected as a person from the moment of his conception.*

Nevertheless, whatever its cause or prognosis, sterility is certainly a difficult trial. The community of believers is called to shed light upon and support the suffering of those who are unable to fulfill their legitimate aspiration to motherhood and fatherhood. Spouses who find themselves in this sad situation are called to find in it an opportunity for sharing in a particular way in the Lord's Cross, the source of spiritual fruitfulness. Sterile couples must not forget that "even when procreation is not possible, conjugal life doe snot for this reason lose its value. Physical sterility in fact can be for spouses the occasion for other important services to the life of the human person, for example, adoption, various forms of educational work, and assistance to other families and to poor or handicapped children".[59]

Many researchers are engaged in the fight against sterility. While fully safeguarding the dignity of human procreation, some have achieved results which previously seemed unattainable. Scientists therefore are to be encouraged to continue their research with the aim of preventing the causes of sterility and of being able to remedy them so that sterile couples will be able to procreate in full respect for their own personal dignity and that of the child to be born.

III. MORAL AND CIVIL LAW

The Values and Moral Obligations that Civil Legislation Must Respect and Sanction in this Matter

The inviolable right to life of every innocent human individual and the rights of the family and of the institution of marriage constitute fundamental moral values, because they concern the natural condition and integral vocation of the human person; at the same time they are constitutive elements of civil society and its order.

For this reason the new technological possibilities which have opened up in the field of biomedicine require the intervention of the political authorities and of the legislator, since an uncontrolled application of such techniques could lead to unforeseeable and damaging consequences for civil society. Recourse to the conscience of each individual and to the self-regulation of researchers cannot be sufficient for ensuring respect for personal rights and public order. If the legislator responsible for the common good were not watchful, he could be deprived of his prerogatives by researchers claiming to govern humanity in the name of the biological discoveries and the alleged "improvement" processes which they would draw from those discoveries. "Eugenism" and forms of discrimination between human beings could come to be legitimized: this would constitute an act of violence and a serious offense to the equality, dignity and fundamental rights of the human person.

The intervention of the public authority must be inspired by the rational principles which regulate the relationships between civil law and moral law. The task of civil law is to ensure the common good of people through the recognition of and the defence of fundamental rights and through the promotion of peace and of public morality.[60] In no sphere of life can the civil law take the place of conscience or dictate norms concerning things which are outside its competence. It must sometimes tolerate, for the sake of public order, things which it cannot forbid without a greater evil resulting. However, the inalienable rights of the person must be recognized and respected by civil society and the political authority. These human rights depend neither on single individuals nor on parents; nor do they represent a concession made by society and the State: they pertain to human nature and are inherent in the person by virtue of the creative act from which the person tool his or her origin.

Among such fundamental rights one should mention in this regard: a) every human being's right to life and physical integrity from the moment of conception until death; b) the rights of the family and of marriage as an institution and, in this area, the child's right to be conceived, brought into this world and brought up by his parents. To each of these two themes it is necessary here to give some further consideration.

In various States certain laws have authorized the direct suppression of innocents: the moment a positive law deprives a category of human beings of the protection which civil legislation must accord them, the State is denying the equality of all before the law. When the State does not place its power at the service of the rights of each citizen, and in particular of the more vulnerable, the very foundations of a State based on law are undermined. The political authority consequently cannot give approval to the calling of human beings into existence through procedures which would expose them to those very grave risks noted previously. The possible recognition by positive law and the political authorities of techniques of artificial transmission of life and the experimentation connected with it would widen the breach already opened by the legalization of abortion.

As a consequence of the respect and protection which must be ensured for the unborn child from the moment of his conception, the law must provide appropriate penal sanctions for every deliberate violation of the child's rights. The law cannot tolerate – indeed it must expressly forbid – that human beings, even at the embryonic stage, should be treated as objects of experimentation, be mutilated or destroyed with the excuse that they are superfluous or incapable of developing normally.

The political authority is bound to guarantee to the institution of the family, upon which society is based, the judicial protection to which it has a right. From the very fact that it is at the service of people, the political authority must also be at the service of the family. Civil law cannot grant approval to techniques of artificial procreation which, for the benefit of third parties (doctors, biologists, economic or governmental powers), take away what is a right inherent in the relationship between spouses; and therefore civil law cannot legalize the donation of gametes between persons who are not legitimately united in marriage.

Legislation must also prohibit, by virtue of the support which is due to the family, embryo banks, *post mortem* insemination and "surrogate motherhood".

It is part of the duty of the public authority to ensure that the civil law is regulated according to the fundamental norms of the moral law in matters concerning human rights, human life and the institution of the family. Politicians must commit themselves, through their interventions upon public opinion, to securing in society the widest possible consensus on such essential points and to consolidating this consensus wherever it risks being weakened or is in danger of collapse.

In many countries, the legalization of abortion and judicial tolerance of unmarried couples makes it more difficult to secure respect for the fundamental rights recalled by this Instruction. It is to be hoped that States will not become responsible for aggravating these socially damaging situations of injustice. It is rather to be hoped that nations and States will realize all the cultural, ideological and political implications connected with the techniques of artificial procreation and will find the wisdom and courage necessary for issuing laws which are more just and respectful to human life and the institution of the family.

The civil legislation of many states confers an undue legitimation upon certain practices in the eyes of many today; it is seen to be incapable of guaranteeing that morality which is in conformity with the natural exigencies of the human person and with the "unwritten laws" etched by the Creator upon the human heart. All men of good will must commit themselves, particularly within their professional field and in the exercise of their civil rights, to ensuring the reform of morally unacceptable civil laws and the correction of illicit practices. In addition, "conscientious objection" vis-a-vis such laws must be supported and recognized. A movement pf passive resistance to the legitimation of practices contrary to human life and dignity is beginning to make an even sharper impression upon the moral conscience of many, especially among specialists in the biomedical sciences.

CONCLUSION

The spread of technologies of intervention in the processes of human procreation raises very serious moral problems in relation to the respect due to the human being from the moment of conception, to the dignity of the person, of his or her sexuality, and of the transmission of life.

With this Instruction the Congregation for the Doctrine of the Faith, in fulfilling its responsibility to promote and defend the Church's teaching in so serious a matter, addresses a new and heartfelt invitation to all those who, by reason of their role and their commitment, can exercise a positive influence and ensure that, in the family and in society, due respect is accorded to life and love. It addresses this invitation to those responsible for the formation of consciences and of public opinion, to scientists and medical professionals, to jurists and politicians. It hopes that all will understand the incompatibility between recognition of the dignity of the human person and contempt for life and love, between faith in the living God and the claim to decide arbitrarily the origin and fate of a human being.

In particular, the Congregation for the Doctrine of the Faith addresses an invitation with confidence and encouragement to theologians, and above all to moralists, that they study more deeply and make ever more accessible to the faithful the contents of the teaching of the Church's Magisterium in the light of a valid anthropology in the matter of sexuality and marriage and in the context of the necessary interdisciplinary approach. Thus they will make it possible to understand ever more clearly the reasons for and the validity of this teaching. By defending man against the excesses of his own power, the Church of God reminds him of the reasons for his true nobility; only in this way can the possibility of living and loving with that dignity and liberty which derive from respect for the truth be ensured for the men and women of tomorrow. The precise indications which are offered in the present Instruction therefore are not meant to halt the effort of reflection but rather to give it a renewed impulse in unrenounceable fidelity to the teaching of the Church.

In the light of the truth about the gift of human life and in the light of the moral principles which flow from that truth, everyone is invited to act in the area of responsibility proper to each and, like the good Samaritan, to recognize as a neighbour even the littlest among the children of men (Cf. *Lk.* 10:29–37). Here Christ's words find a new and particular echo:

"What you do to one of the least of my brethren, you do unto me" (*Mt* 25:40).

During an audience granted to the undersigned Prefect after the plenary session of the Congregation for the Doctrine of the Faith, the Supreme Pontiff, John Paul II, approved this Instruction and ordered it to be published.

Given at Rome, from the Congregation for the Doctrine of the Faith, February 22, 1987, the Feast of the Chair of St. Peter, the Apostle.

United States Catholic Conference,
March 1987
Reprinted with permission of USCC

JOSEPH CARD. RATZINGER
Prefect
ALBERTO BOVONE
Titular Archibishop of Caesarea in Numidia Secretary

NOTES

[1] Pope John Paul II, *Discourse to those taking part in the 81st Congress of the Italian Society of International Medicine and the 82nd Congress of the Italian Society of General Surgery*, 27 October 1980: AAS 72 (1980) 1126.

[2] Pope John Paul VI, *Discourse to the General Assembly of the United Nations Organization*, 4 October 1965: AAS 57 (1965) 878; Encyclical *Populorum Progressio*, 13: AAS 59 (1967) 263.

[3] Pope John Paul VI, *Homily during the Mass closing the Holy Year*, 25 December 1975: AAS 68 (1976) 145; Pope John Paul II, Encyclical *Dives in Misericordia*, 30; AAS 72 (1980) 1224.

[4] Pope John Paul II, *Discourse to those taking part in the 35th General Assembly of the World Medical Association*, 29 October 1983: AAS 76 (1983) 390.

[5] Cf. Declaration *Dignitatis Humanae*, 2.

[6] Pastoral Constitution *Gaudium et Spes*, 22; Pope John Paul II, Encyclical *Redemptor Hominis*, 8: AAS 71 (1979) 270–272.

[7] Cf. Pastoral Constitution *Gaudium et Spes*, 35.

[8] Pastoral Constitution *Gaudium et Spes*, 15; cf. Also Pope Paul VI, Encyclical *Populorum Progressio*, 20: AAS 59 (1967) 267; Pope John Paul II, Encyclical *Redemptor Hominis*, 15: AAS 71 (1979) 286–289; Apostolic Exhortation *Familiaris Consortio*, 8: AAS 74 (1982) 89.

[9] Pope John Paul II, Apostolic Exhortation *Familiaris Consortio*, 11: AAS 74 (1982) 92.

[10] Cf. Pope Paul VI, Encyclical *Humanae Vitae*, 10: AAS 60 (1968) 487–488.

[11] Pope John Paul II, *Discourse to the members of the 35th General Assembly of the World Medical Association*, 29 October 1983: AAS 76 (1984) 393.

[12] Cf. Pope John Paul II, Apostolic Exhortation *Familiaris Consortio*, 11: AAS 74 (1982) 91–92; cf. Also Pastoral Constitution *Gaudium et Spes*, 50.

[13] Sacred Congregation for the Doctrine of the Faith, *Declaration on Procured Abortion*, 9, AAS (1974) 736–737.
[14] Pope John Paul II, *Discourse to those taking part in the 35th General Assembly of the World Medical Association*, 29 October 1983: AAS 76 (1984) 390.
[15] Pope John XXIII, Encyclical *Mater et Magistra*, III: AAS 53 (1961) 447.
[16] Pastoral Constitution *Gaudium et Spes*, 24.
[17] Cf. Pope Pius XII, Encyclical *Humani Generis*: AAS 42 (1950) 575; Pope Paul VI, *Professio Fidei*: AAS 60 (1968) 436.
[18] Pope John XXIII, Encyclical *Mater et Magistra*, III: AAS 53 (1961) 447; cf. Pope John Paul II, *Discourse to priests participating in a seminar on "Responsible Procreation"*, 17 September 1983, *Insegnamenti di Giovanni Paolo II*, VI, 2 (1983) 562: "At the origin of each human person there is a creative act of God: no man comes into existence by chance; he is always the result of the creative love of God".
[19] Cf. Pastoral Constitution *Gaudium et Spes*, 24.
[20] Cf. Pope Pius XII, *Discourse to the Saint Luke Medical-Biological Union*, 12 November 1944: *Discorsi e Radiomessaggi* VI (1944–1945) 191–192.
[21] Cf. Pastoral Constitution *Gaudium et Spes*, 50.
[22] Cf. Pastoral Constitution *Gaudium et Spes*, 51: "When it is a question of harmonizing married love with the responsible transmission of life, the moral character of one's behaviour does not depend only on the good intention and the evaluation of the motives: the objective criteria must be used, criteria drawn from the nature of the human person and human acts, criteria which respect the total meaning of mutual self-giving and human procreation in the context of true love".
[23] Pastoral Constitution *Gaudium et Spes*, 51.
[24] Holy See, *Charter of the Rights of the Family*, 4: *L'Osservatore Romano*, 25 November 1983.
[25] Sacred Congregation for the Doctrine of the Faith, *Declaration of Procured Abortion, 12–13: AAS 66 (1974) 738.*
[26] Cf. Pope Paul VI, *Discourse to participants in the Twenty-third National Congress of Italian Catholic Jurists*, 9 December 1972: AAS 64 (1972) 777.
[27] The obligation to avoid disproportionate risks involves an authentic respect for human beings and the uprightness of therapeutic intentions. It implies that the doctor "above all ...must carefully evaluate the possible negative consequences which the necessary use of a particular exploratory technique may have upon the unborn child and avoid recourse to diagnostic procedures which do not offer sufficient guarantees of their honest purpose and substantial harmlessness. And if, as often happens in human choices, a degree of risk must be undertaken, he will take care to assure that it is justified by a truly urgent need for the diagnosis and by the importance of the results that can be achieved by it for the benefit of the unborn child himself" (Pope John Paul II, *Discourse to Participants in the Pro-Life Movement Congress*, 3 December 1982: *Insegnamenti di Giovanni Paolo II*, 3 [1982] 1512). This clarification concerning "proportionate risk" is also to be kept in mind in the following sections of the present Instruction, whenever this term appears.
[28] Pope John Paul II, *Discourse to the Participants in the 35th General Assembly of the World Medical Association*, 29 October 1983: AAS 76 (1984) 392.
[29] Cf. Pope John Paul II, *Address to a Meeting of the Pontifical Academy of Sciences*, 23 October 1982: AAS 75 (1983) 37: "I condemn, in the most explicit and formal

way, experimental manipulations of the human embryo, since the human beings from conception to death, cannot be exploited for any purpose whatsoever".

[30] Holy See, *Charter of the Rights of the Family*, 4b: *L'Osservatore Romano*, 25 November 1983.

[31] Cf. Pope John Paul II, *Address to the Participants in the Convention of the Pro-Life Movement*, 3 December 1982: *Insegnamenti di Giovanni Paolo II*, V, 3 (1982) 1511: "Any form of experimentation on the foetus that may damage its integrity or worsen its condition is unacceptable, except in the case of a final effort to save it from death". Sacred Congregation for the Doctrine of the Faith, *Declaration on Euthanasia*, 4: AAS 72 (1980) 550: "In the absence of other sufficient remedies, it is permitted, with the patient's consent, to have recourse to the means provided by the most advanced medical techniques, even if these means are still at the experimental stage and are not without a certain risk".

[32] No one before coming into existence, can claim a subjective right to begin to exist; nevertheless, it is legitimate to affirm the right of the child to have a fully human origin through conception in conformity with the personal nature of the human being. Life is a gift that must be bestowed in a manner worthy of the both the subject receiving it and of the subjects transmitting it. This statement is to be borne in mind for what will be explained concerning artificial human procreation.

[33] Cf. Pope John Paul II, *Discourse to those taking part in the 35th General Assembly of the World Medical Association*, 29 October 1983: AAS 76 (1984) 391.

[34] Cf. Pastoral Constitution on the Church in the Modern World, *Gaudium et Spes*, 50.

[35] Cf. Pope John Paul II, Apostolic Exhortation *Familiaris Consortio*, 14: AAS 74 (1982) 96.

[36] Cf. Pope Pius XII, *Discourse to those taking part in the 4th International Congress of Catholic Doctors*, 29 September 1949: AAS 41 (1949) 559. According to the plan of the creator, "A man leaves his father and his mother and cleaves to his wife, and they become one flesh" (*Gen* 2:24). The unity of marriage, bound to the order of creation, is a truth accessible to natural reason. The Church's Tradition and Magisterium frequently make reference to the Book of Genesis, both directly and through the passages of the New Testament that refer to it: *Mt* 19:4–6; *Mk* 10:5–8; *Eph* 5:31. Cf. Athenagoras, *Legatio pro christianis*, 33: PG 6, 965–967; St Chrysostom, *In Matthaeum homilae*, LXII, 19, 1: PG 58 597; St Leo the Great, *Epist. Ad Rusticum*, 4: PL 54, 1204; Innocent III, Epist. *Gaudemus in Domino*: DS 778; Council of Lyons II, *IV Session*: DS 860; Council of Trent, *XXIV Session*: DS 1798. 1802; Pope LEO XIII, Encyclical *Arcanum Divinae Sapientiae*: ASS 12 (1879/80) 388–391; Pope Pius XI, Encyclical *Casti Connubii*: AAS 22 (1930) 546–547; Second Vatican Council, *Gaudium et Spes*, 48; Pope John Paul II, Apostolic Exhortation *Familiaris Consortio*, 19: AAS 74 (1982) 101–102; *Code of Canon Law*, Can. 1056.

[37] Cf. Pope Pius XII, *Discourse to those taking part in the 4th International Congress of Catholic Doctors*, 29 September 1949: AAS 41 (1949) 560; *Discourse to those taking part in the congress of the Italian Catholic Union of Midwives*, 29 October 1951: AAS 43 (1951) 850; *Code of Canon Law*, Can. 1134.

[38] Pope Paul VI, Encyclical Letter *Humanae Vitae*, 12: AAS 60 (1968) 488–489.

[39] *Loc. cit., ibid.*, 489.

[40] Pope Pius XII, *Discourse to those taking part in the Second Naples World Congress on Fertility and Human Sterility*, 19 May 1956: AAS 48 (1956) 470.

[41] *Code of Canon Law*, Can. 1061. According to this Canon, the conjugal act is that by which the marriage is consummated if the couple "have performed (it) between themselves in a human manner".

[42] Cf. Pastoral Constitution *Gaudium et Spes*, 14.

[43] Cf. Pope John Paul II, General Audience on 16 January 1980: *Insegnamenti di Giovanni Paolo II*, III, 1 (1980) 148–152.

[44] Pope John Paul II, *Discourse to those taking part in the 35th General Assembly of the World Medical Association*, 29 October 1983: AAS 76 (1984) 393.

[45] Cf. Pastoral Constitution *Gaudium et Spes*, 51.

[46] Cf. Pastoral Constitution *Gaudium et Spes*, 50.

[47] Cf. Pope Pius XII, *Discourse to those taking part in the 4th International congress of Catholic Doctors*, 29 September 1949: AAS 41 (1949) 560: "It would be erroneous ...to think that the possibility of resorting to this means (artificial fertilization) might render valid a marriage between persons unable to contract it because of the *impedimentum impotentiae*".

[48] A similar question was dealt with by Pope Paul VI, Encyclical *Humanae Vitae*, 14: AAS 60 (1968) 490–491.

[49] Cf. *Supra*: I, 1 ff.

[50] Pope Paul II, Apostolic Exhortation *Familiaris Consortio*, 14: AAS 74 (1982) 96.

[51] Cf. *Response of the Holy Office*, 17 March 1897: DS 3323; Pope Pius XII, *Discourse to those taking part in the 4th International Congress of Catholic Doctors*, 29 September 1949: AAS 41 (1949) 560; *Discourse to the Italian Catholic Union of Midwives*, 29 October 1951: AAS 43 (1951) 850; *Discourse to those taking part in the Second Naples World Congress on Fertility and Human Sterility*, 19 May 1956: AAS 49 (1956) 471–473; *Discourse to those taking part in the 7th International Congress of the International Society of Haematology*, 12 September 1958: AAS 50 (1958) 733; Pope John XXIII, Encyclical *Mater et Magistra*, III: AAS 53 (1961) 447.

[52] Pope Pius XII, *Discourse to the Italian Catholic Union of Midwives*, 29 October 1951: AAS 43 (1951) 850.

[53] Pope Pius XII, *Discourse to those taking part in the 4th International Congress of Catholic Doctors*, 29 September 1949: AAS 41 (1949) 560.

[54] Sacred Congregation for the Doctrine of the Faith, *Declaration on Certain Questions concerning Sexual Ethics*, 9: AAS 68 (1976) 86, which quotes the Pastoral Constitution *Gaudium et Spes*, 51. Cf. *Decree of the Holy Office*, 2 August 1929: AAS 21 (1929) 490; Pope Pius XII, *Discourse to those taking part in the 26th Congress of the Italian Society of Urology*, 8 October 1953: AAS 45 (1953) 678.

[55] Cf. Pope John XXIII, Encyclical *Mater et Magistra*, III: AAS 53 (1961) 447.

[56] Cf. Pope Pius XII, *Discourse to those taking part in the 4th International Congress of Catholic Doctors*, 29 September 1949: AAS 41 (1949), 560.

[57] Cf. Pope Pius XII, *Discourse to those taking part in the Second Naples World Congress on Fertility and Human Sterility*, 19 May 1956: AAS 48 (1956) 471–473).

[58] Pastoral constitution *Gaudium et Spes*, 50.

[59] Pope John Paul II, Apostolic Exhortation *Familiaris Consortio*, 14: AAS 74 (1982) 97.

[60] Cf. Declaration *Dignitatis Humanae*, 7.

NOTES ON CONTRIBUTORS

DEBORAH D. BLAKE, Ph.D., Department of Religious Studies, Regis University, Denver, CO, 80221, U.S.A.
JOHN W. CARLSON, Ph.D., Department of Philosophy, Creighton University, Omaha, NE, 68178, U.S.A.
PETER J. FAGAN, Ph.D., Director, Consultation Unit, A Johns Hopkins Medical Institution, 550 North Broadway, Baltimore, MD, 21205, U.S.A.
PATRICIA B. JUNG, Ph.D., Wartburg Theological Seminary, Dubuque, IA, 52003, U.S.A.
RON HAMEL, Ph.D., Director, Department of Clinical Ethics, Lutheran General Hospital, Park Ridge, IL, 60068, U.S.A.
JAMES F. KEENAN, S.J., Weston Jesuit School of Theology, Cambridge, MA, 02138, U.S.A.
WILLIAM E. MAY, John Paul II Institute for Studies on Marriage and Family, 487 Michigan Avenue, NE, Washington, DC, 20017, U.S.A.
YULA PONTICAS, Ph.D., Senior Staff Psychologist, Consultation Unit, A Johns Hopkins Medical Institution, 550 North Broadway, Baltimore, MD, 21205, U.S.A.
JEAN PORTER, Ph.D., The Department of Theology, The University of Notre Dame, Notre Dame, IN, 46556, U.S.A.
MIRIAM B. ROSENTHAL, M.D., Psychiatry and Reproductive Biology, Case Western Reserve University, School of Medicine, Cleveland, OH, 44106, U.S.A.
JAMES A. SIMON, M.D., Women's Health Research Center, 14201 Laurel Park Drive, Suite #104, Laurel, MD, 20707, U.S.A.
CAROL A. TAUER, Ph.D., Philosophy Department, The College of St. Catherine, St. Paul, MN, 55105, U.S.A.
KEVIN WM. WILDES, S.J., Ph.D., Department of Philosophy, The Kennedy Institute of Ethics, The Center for Clinical Bioethics, Georgetown University, Washington, DC, 20057, U.S.A.

INDEX

Andrews, F. 152, 153
Andrews, L. 170
anthropology 211–213
Aquinas, T. 53, 55, 56, 59, 61, 64, 107, 121, 183, 186, 188, 199
Ashley, B. 85

Baruch, E. 150
Berg, B. 151, 152
Blackstone, W. 181
Blake, D. 4, 149–166, 198, 203, 205, 206
Brody, B. 133
Brown, Louise 1, 9, 73, 128
Burns, L. 40, 41, 48

Cahill, L. 2, 60, 63, 75, 86, 130, 133, 149, 155, 156
Carlson, J. 3, 58, 68, 107–124, 201, 202
Center for Disease Control 39
Charter of the Rights of the Family 215
common good 162
Congregation for the Doctrine of the Faith (CDF) 2, 58, 73, 76, 77, 93, 95, 96, 107, 108, 117, 130, 149, 171, 173, 174, 209–238
consensus 132–133
consultation 130–132
Curran, C. 127, 128, 134, 136

Daniel, W. 138
Davis, H. 66, 68
Declaration on Procured Abortion 215

Demmer, K. 65, 68
Denzinger 57, 68
Devlin, P. 181, 190
Donum Vitae 2, 29, 53, 61, 62, 63, 65, 73, 76, 77, 93, 94, 95, 99, 100, 102, 104, 107, 108, 109, 117, 118–121, 125, 127, 129, 130, 131, 132, 133, 134, 135, 138, 139, 143, 144, 171, 181, 182, 193, 197, 199, 200, 201, 202, 205, 209–238
Downey, J. 42, 48

embryo development 10
embryo research and experimentation 218–220
Englehardt, H. T. 186
Ericson, E. 41

Fagen, P. 2, 3, 27–38, 65, 69, 201, 203
Falise, M. 129
Finnis, J. 81
Freeman, E. W. 28, 38
Fuchs, J. 54

Gallagher, J. 53, 69
Ghoos, J. 59, 69
GIFT 113–121
Greenfeld, G. 43, 48
Greil, A. L. 153, 167
Griese, O. 109
Griffin, L. 66, 69

Hammel, R. 4, 197–208
Harvey, J. 53, 63, 130, 132

Higgins K. L. 40, 48
Humanae Vitae 73, 75, 93, 96, 104, 135, 136, 138, 143
infertility, psychology of 39–49
inseparability principle 3, 75–78, 95–99, 135–139, 198
Instruction on Respect for Human Life in its Origin and on the Dignity of Procreation, see *Donum Vitae*
IVF, history 11–12
IVF, psychological evaluation 27–38

Janssens, L. 60, 69
Jersild, P. 171
John XXIII 161, 162
John Paul II, 2, 77, 108
Johns Hopkins Hospital 28, 29
Jonsen, A. and Toulmin, S. 60, 69
Jung, P. 4, 167–180, 203, 205, 206

kantian 100–104
Kass, L. 88
Keane 54
Keenan, J. 2, 53–71, 186, 199
Kelly, G. 66, 69
Kiely, B. 60, 69
Klawiter, M. 150
Kopfensteiner, T. 55
Kraft, A. 44, 48

Leo XIII 161, 162

MacIntyre 54, 63
magisterium 209–238
Mahoney, J. 60, 70
Mangan, J. 59, 70
marital rights 78–81
manualist 54–61
May, W. 3, 60, 70, 73–92, 99, 100, 104, 105, 109, 207

McCormick, R. 2, 60, 63, 75, 81–87, 94, 97, 98, 103, 104, 105, 128, 129, 130, 131, 132, 136, 139, 143
Menning, B. E. 41, 42, 48
Morse, C. 28, 38
moral truth, subjective-objective 54ff
moral object 53–71
Mullady, B. 55, 70

natural law 182–185
Nelson, D. 67, 70

O'Donnell, T. 128
Office of Technology Assessment (OTA) 110, 155, 156
Olshansky, E. F. 40, 48

Paul VI 73, 75, 80, 108, 162, 163
Pius XII 2, 58, 75, 80, 97, 103, 108, 132, 149
Ponticas, Y. 2, 3, 27–38, 201, 203
Porter, J. 3, 93–106, 200, 207
Post, S. 66, 70
psychological needs 151–154
prenatal diagnosis 216–217
public policy 156–160

Ratzinger, C. 129, 130
reproductive technologies 12–22
resource allocation 167–180
Rosenthal 2, 3, 39–49, 203

sacramental theology and marriage 143–145
Seibel, M. 39, 49
Shannon 86, 149
Simon, J. 2, 3, 9–25
Smith, J. 87
social context 154–156
social responsibility 160
solidarity 162–163

Speroff, L. 39, 42, 49
state authority 181–194

Tauer, C. 4, 125–146, 197, 199, 200, 201, 203, 206
Thielicke, H. 78
toleration 65ff
TOTS, 113–121
Trau, J. 63, 70
Treloar, J. 54, 70

Ulshafer, T. 54, 66, 71

Vacek, E. 60, 63, 71, 130, 132, 139
Valsecchi, A. 57, 71
Vaux, K. 93
Vermeersch, A. 128
Visser, J. 128

Walters, L. 129, 130, 133, 168
Wildes, K. 1–5, 181–194, 202
Williams, L. P. 173, 178

Zalba, M. 128

Philosophy and Medicine

1. H. Tristram Engelhardt, Jr. and S.F. Spicker (eds.): *Evaluation and Explanation in the Biomedical Sciences.* 1975   ISBN 90-277-0553-4
2. S.F. Spicker and H. Tristram Engelhardt, Jr. (eds.): *Philosophical Dimensions of the Neuro-Medical Sciences.* 1976   ISBN 90-277-0672-7
3. S.F. Spicker and H. Tristram Engelhardt, Jr. (eds.): *Philosophical Medical Ethics.* Its Nature and Significance. 1977   ISBN 90-277-0772-3
4. H. Tristram Engelhardt, Jr. and S.F. Spicker (eds.): *Mental Health.* Philosophical Perspectives. 1978   ISBN 90-277-0828-2
5. B.A. Brody and H. Tristram Engelhardt, Jr. (eds.): *Mental Illness.* Law and Public Policy. 1980   ISBN 90-277-1057-0
6. H. Tristram Engelhardt, Jr., S.F. Spicker and B. Towers (eds.): *Clinical Judgment.* A Critical Appraisal. 1979   ISBN 90-277-0952-1
7. S.F. Spicker (ed.): *Organism, Medicine, and Metaphysics.* Essays in Honor of Hans Jonas on His 75th Birthday. 1978   ISBN 90-277-0823-1
8. E.E. Shelp (ed.): *Justice and Health Care.* 1981   ISBN 90-277-1207-7; Pb 90-277-1251-4
9. S.F. Spicker, J.M. Healey, Jr. and H. Tristram Engelhardt, Jr. (eds.): *The Law-Medicine Relation.* A Philosophical Exploration. 1981   ISBN 90-277-1217-4
10. W.B. Bondeson, H. Tristram Engelhardt, Jr., S.F. Spicker and J.M. White, Jr. (eds.): *New Knowledge in the Biomedical Sciences.* Some Moral Implications of Its Acquisition, Possession, and Use. 1982   ISBN 90-277-1319-7
11. E.E. Shelp (ed.): *Beneficence and Health Care.* 1982   ISBN 90-277-1377-4
12. G.J. Agich (ed.): *Responsibility in Health Care.* 1982   ISBN 90-277-1417-7
13. W.B. Bondeson, H. Tristram Engelhardt, Jr., S.F. Spicker and D.H. Winship: *Abortion and the Status of the Fetus.* 2nd printing, 1984   ISBN 90-277-1493-2
14. E.E. Shelp (ed.): *The Clinical Encounter.* The Moral Fabric of the Patient-Physician Relationship. 1983   ISBN 90-277-1593-9
15. L. Kopelman and J.C. Moskop (eds.): *Ethics and Mental Retardation.* 1984   ISBN 90-277-1630-7
16. L. Nordenfelt and B.I.B. Lindahl (eds.): *Health, Disease, and Causal Explanations in Medicine.* 1984   ISBN 90-277-1660-9
17. E.E. Shelp (ed.): *Virtue and Medicine.* Explorations in the Character of Medicine. 1985   ISBN 90-277-1808-3
18. P. Carrick: *Medical Ethics in Antiquity.* Philosophical Perspectives on Abortion and Euthanasia. 1985   ISBN 90-277-1825-3; Pb 90-277-1915-2
19. J.C. Moskop and L. Kopelman (eds.): *Ethics and Critical Care Medicine.* 1985   ISBN 90-277-1820-2
20. E.E. Shelp (ed.): *Theology and Bioethics.* Exploring the Foundations and Frontiers. 1985   ISBN 90-277-1857-1
21. G.J. Agich and C.E. Begley (eds.): *The Price of Health.* 1986   ISBN 90-277-2285-4
22. E.E. Shelp (ed.): *Sexuality and Medicine.* Vol. I: Conceptual Roots. 1987   ISBN 90-277-2290-0; Pb 90-277-2386-9

Philosophy and Medicine

23. E.E. Shelp (ed.): *Sexuality and Medicine.* Vol. II: Ethical Viewpoints in Transition. 1987   ISBN 1-55608-013-1; Pb 1-55608-016-6
24. R.C. McMillan, H. Tristram Engelhardt, Jr., and S.F. Spicker (eds.): *Euthanasia and the Newborn.* Conflicts Regarding Saving Lives. 1987   ISBN 90-277-2299-4; Pb 1-55608-039-5
25. S.F. Spicker, S.R. Ingman and I.R. Lawson (eds.): *Ethical Dimensions of Geriatric Care.* Value Conflicts for the 21th Century. 1987   ISBN 1-55608-027-1
26. L. Nordenfelt: *On the Nature of Health.* An Action-Theoretic Approach. 2nd, rev. ed. 1995   ISBN 0-7923-3369-1; Pb 0-7923-3470-1
27. S.F. Spicker, W.B. Bondeson and H. Tristram Engelhardt, Jr. (eds.): *The Contraceptive Ethos.* Reproductive Rights and Responsibilities. 1987   ISBN 1-55608-035-2
28. S.F. Spicker, I. Alon, A. de Vries and H. Tristram Engelhardt, Jr. (eds.): *The Use of Human Beings in Research.* With Special Reference to Clinical Trials. 1988   ISBN 1-55608-043-3
29. N.M.P. King, L.R. Churchill and A.W. Cross (eds.): *The Physician as Captain of the Ship.* A Critical Reappraisal. 1988   ISBN 1-55608-044-1
30. H.-M. Sass and R.U. Massey (eds.): *Health Care Systems.* Moral Conflicts in European and American Public Policy. 1988   ISBN 1-55608-045-X
31. R.M. Zaner (ed.): *Death: Beyond Whole-Brain Criteria.* 1988   ISBN 1-55608-053-0
32. B.A. Brody (ed.): *Moral Theory and Moral Judgments in Medical Ethics.* 1988   ISBN 1-55608-060-3
33. L.M. Kopelman and J.C. Moskop (eds.): *Children and Health Care.* Moral and Social Issues. 1989   ISBN 1-55608-078-6
34. E.D. Pellegrino, J.P. Langan and J. Collins Harvey (eds.): *Catholic Perspectives on Medical Morals.* Foundational Issues. 1989   ISBN 1-55608-083-2
35. B.A. Brody (ed.): *Suicide and Euthanasia.* Historical and Contemporary Themes. 1989   ISBN 0-7923-0106-4
36. H.A.M.J. ten Have, G.K. Kimsma and S.F. Spicker (eds.): *The Growth of Medical Knowledge.* 1990   ISBN 0-7923-0736-4
37. I. Löwy (ed.): *The Polish School of Philosophy of Medicine.* From Tytus Chałubiński (1820–1889) to Ludwik Fleck (1896–1961). 1990   ISBN 0-7923-0958-8
38. T.J. Bole III and W.B. Bondeson: *Rights to Health Care.* 1991   ISBN 0-7923-1137-X
39. M.A.G. Cutter and E.E. Shelp (eds.): *Competency.* A Study of Informal Competency Determinations in Primary Care. 1991   ISBN 0-7923-1304-6
40. J.L. Peset and D. Gracia (eds.): *The Ethics of Diagnosis.* 1992   ISBN 0-7923-1544-8

Philosophy and Medicine

41. K.W. Wildes, S.J., F. Abel, S.J. and J.C. Harvey (eds.): *Birth, Suffering, and Death*. Catholic Perspectives at the Edges of Life. 1992 [CSiB-1]
 ISBN 0-7923-1547-2; Pb 0-7923-2545-1
42. S.K. Toombs: *The Meaning of Illness*. A Phenomenological Account of the Different Perspectives of Physician and Patient. 1992
 ISBN 0-7923-1570-7; Pb 0-7923-2443-9
43. D. Leder (ed.): *The Body in Medical Thought and Practice*. 1992
 ISBN 0-7923-1657-6
44. C. Delkeskamp-Hayes and M.A.G. Cutter (eds.): *Science, Technology, and the Art of Medicine*. European-American Dialogues. 1993   ISBN 0-7923-1869-2
45. R. Baker, D. Porter and R. Porter (eds.): *The Codification of Medical Morality*. Historical and Philosophical Studies of the Formalization of Western Medical Morality in the 18th and 19th Centuries, Volume One: Medical Ethics and Etiquette in the 18th Century. 1993   ISBN 0-7923-1921-4
46. K. Bayertz (ed.): *The Concept of Moral Consensus*. The Case of Technological Interventions in Human Reproduction. 1994   ISBN 0-7923-2615-6
47. L. Nordenfelt (ed.): *Concepts and Measurement of Quality of Life in Health Care*. 1994 [ESiP-1]   ISBN 0-7923-2824-8
48. R. Baker and M.A. Strosberg (eds.) with the assistance of J. Bynum: *Legislating Medical Ethics*. A Study of the New York State Do-Not-Resuscitate Law. 1995   ISBN 0-7923-2995-3
49. R. Baker (ed.): *The Codification of Medical Morality*. Historical and Philosophical Studies of the Formalization of Western Morality in the 18th and 19th Centuries, Volume Two: Anglo-American Medical Ethics and Medical Jurisprudence in the 19th Century. 1995   ISBN 0-7923-3528-7; Pb 0-7923-3529-5
50. R.A. Carson and C.R. Burns (eds.): *The Philosophy of Medicine and Bioethics*. Retrospective and Critical Appraisal. (forthcoming)   ISBN 0-7923-3545-7
51. K.W. Wildes, S.J. (ed.): *Critical Choices and Critical Care*. Catholic Perspectives on Allocating Resources in Intensive Care Medicine. 1995 [CSiB-2]
 ISBN 0-7923-3382-9
52. K. Bayertz (ed.): *Sanctity of Life and Human Dignity*. 1996
 ISBN 0-7923-3739-5
53. Kevin Wm. Wildes, S.J. (ed.): *Infertility: A Crossroad of Faith, Medicine, and Technology*. 1996   ISBN 0-7923-4061-2
54. Kazumasa Hoshino (ed.): *Japanese and Western Bioethics*. Studies in Moral Diversity. 1996   ISBN 0-7923-4112-0